Multinationals, Governments and International Technology Transfer

New inventions, original products, technical processes and technical experience are among the technological assets of a firm and they constitute one of the main sources of its competitive power. In transferring technology in order to exploit international markets, multinationals seek to maximise the return on their technological advantages and retain their competitive edge. At the same time, many host-country governments are keen that technology transfer should be on terms which benefit their national interests, hence their frequent preference for multinationals to engage in joint ventures with, or license their technology to, local companies. This book examines the international technology transfer process and the role of both multinationals and host-country governments in that process, with emphasis on the experience of the more developed countries. It explores a range of issues and presents much original thinking and research findings. It discusses in particular the strategies of multinationals, assessing how far they are willing to accept technology transfer to external partners (as opposed to subsidiaries which they can control). It also examines how far technical transfers are successful from the viewpoint of the firm and countries involved, arguing that governments are most likely to succeed in attracting multinational transfers if they are aware of and accommodate to some degree multinationals' preferences.

A.E. Safarian is Professor of Economics at the University of Toronto, Canada.

Gilles Y. Bertin is Research Director of AREPIT, Paris, France.

The Croom Helm Series in International Business

Academic Editor: Alan M. Rugman, Dalhousie University

Multinationals and Transfer Pricing
Alan M. Rugman and Lorraine Eden

Multinationals: The Swedish Case
Eric Hörnell and Jan-Erik Vahhe

Japanese Participation in British Industry: Trojan Horse or Catalyst for Growth?
John H. Dunning

Managing the Multinational Subsidiary
Hamid Etemad and Louise Séguin Dulude

Multinationals, Governments and International Technology Transfer

Edited by A.E. Safarian and Gilles Y. Bertin

CROOM HELM
London & Sydney

Croom Helm Ltd, Provident House, Burrell Row,
Beckenham, Kent BR3 1AT

Croom Helm Australia, 44–50 Waterloo Road,
North Ryde, 2113, New South Wales

British Library Cataloguing in Publication Data

Multinationals, governments and international
technology transfer. — (Croom Helm series
in international business)
1. Technology transfer 2. International
business enterprises
I. Safarian, A.E. II. Bertin, Gilles Y.
338.9′26 T174.3

ISBN 0–7099–4453–5

Filmset by Mayhew Typesetting, Bristol, England
Printed and bound in Great Britain

Contents

List of Figures and Tables
Notes on Contributors
Preface

1. An Interpretive Summary *A.E. Safarian* 1

Part One: Technology Transfers and the Multinational Enterprise

2. Multinational Enterprises, International Investments and Transfers of Technology: the Elements of an Integrated Approach *Jean-Louis Mucchielli* 11
3. Trends in Technological Competitiveness within the OECD, 1970–80 *Bernadette Madeuf* 34
4. International Knowledge Transfers and Competitiveness: Canada as a Case Study *D.C. MacCharles* 53

Part Two: Multinational Enterprise Strategies and National Policies

5. Contractual Agreements and International Technology Transfers: the Empirical Studies *Bernard Bonin* 73
6. Multinational Enterprises: Transfer Partners and Transfer Policies *Gilles Y. Bertin* 85
7. The Development of Technology in MNEs: a Cross-Country and Industry Study *Hamid Etemad and Louise Séguin Dulude* 101
8. Canadian Foreign Direct Investment *Alan M. Rugman* 120
9. The Timing, Mode and Terms of Technology Transfer: Some Recent Findings *D.G. McFetridge* 135

Part Three: Industry Studies

10. The Protection of Intellectual Property: Pharmaceutical Products in Canada *H.C. Eastman* 153
11. International Investment, Protection and Technical Transfer: a Preliminary Examination of the Franco-Japanese Case *Patrick A. Messerlin* 167
12. Technology Transfers in the Automotive Equipment Industry: the French Case *Daniel Soulié* 180
13. The Japanese Productivity Advantage in Automobile Production: Can it be Transferred to North America? *Melvyn Fuss and Leonard Waverman* 191

Bibliography 207
Author Index 219
Subject Index 221

Figures and Tables

FIGURES

3.1 Technological balance of payments for eight OECD countries 38
6.1 Transfer preferences for an industry 92
6.2 Transfer strategies by industry 95
7.1 A TOBIT analysis of patented foreign inventions and the MNEs' overall patenting activity 110
7.2 A TOBIT analysis of MNEs' share of foreign inventive activity and MNEs' share of foreign employment 115

TABLES

2.1 Determining factors in relocation in the case of differences in technology 24
2.2 Differences in factor endowment, discrepancies, competitive and comparative advantages and relocation 28
3.1 Technological balance of payments 40
3.2 Index of technological dependency, 1969 and 1979 42
3.3 Index of technological competitiveness, 1969 and 1979 42
3.4 Share of income and payments corresponding to flows between affiliated companies 44
3.5 Intensity of R&D expenditure in the OECD area 46
3.6 Average weights of the manufacturing industries of the OECD area (11 countries) during the period 1970–80 47
3.7 Weights of high R&D-intensity industries in total manufacturing output and exports 48
3.8 Specialisation by 'niches', industry specialisation and apparent comparative advantage in higher R&D-intensity industries 49
4.1 Imports, exports and shipments of manufactured goods by manufacturers by sector of control and trade category of enterprises, 1979 61
4.2 Selected comparisons between sectors of control by plant size, manufacturing sector, 1974 63
6.1 Reactions of transfer partners 88
6.2 Dominant transfer strategies 91

6.3 Number of responses on technology transfers 93
6.4 Importance given to transfer as implied by patenting 93
6.5 Ranking by percentage of positive answers 94
6.6 Internal versus external transfer strategies 97
6.7 Summary of multinational transfer strategies 99
7.1 MNEs' inventive activity and characteristics 104
7.2 Share of patented foreign inventions in the MNEs' overall patenting activity 108
7.3 Comparison between MNEs' share of foreign inventive activity and MNEs' share of foreign employment 112
8.1 Canadian foreign direct investment flows, 1960–84 123
8.2 Bilateral stocks of foreign direct investment by Canada and the United States, 1975–84 125
8.3 Value and shares of foreign direct investment in the United States, 1984 and 1975 126
8.4 The largest Canadian industrial multinationals 131
9.1 Transfers abroad of 406 innovations and 548 imitations of 57 US-based multinationals classified by period of US introduction 137
9.2 The relationship between age and transfer order 138
9.3 Mean international transfer lag (1960–79) 140
9.4 National position in the transfer order: Canada and France by time period 142
9.5 National characteristics and policies and the international transfer order 144
9.6 Estimates of a simultaneous transfer mode and order model 148
11.1 French imports from Japan and the NICs, 1973–84 170
11.2 French restrictive measures against Japanese imports 171
11.3 Japanese investments in France, 1968–84 173
11.4 Industrial environment of Japanese investments, 1980 175
12.1 International trade in high-technology products 183
12.2 Patent applications filed in France 185
12.3 Major French applicants for patents 185
12.4 Major foreign applicants for patents 186
12.5 Technology transfers, French automobile industry, 1973–83 188
13.1 Comparative costs and labour productivity in selected United States and Japanese automobile companies 194
13.2 Comparative costs and labour productivity in selected United States and Japanese automobile companies, 1981 197

13.3 Decomposition of Japanese cost advantage 1983 (Ford-Toyo Kogyo) 197
13.4 Hypothetical sources of Japanese unit cost advantage: US assembly and stamping 200

Notes on Contributors

Gilles Y. Bertin	Université de Paris — Dauphine, CNRS, Directeur de l'AREPIT
Bernard Bonin	Ecole nationale d'administration publique, Université du Québec, Montréal
Louise Séguin Dulude	Centre d'études en administration internationale, Ecole des Hautes Etudes Commerciales de Montréal
H.C. Eastman	Department of Economics, University of Toronto
Hamid Etemad	Faculty of Management, McGill University, Montréal
Melvyn Fuss	Department of Economics University of Toronto
Bernadette Madeuf	University of Limoges, LAREA-CEREM, Université de Paris X-Nanterre
D.C. MacCharles	Division of Social Science, University of New Brunswick, Saint John
D.G. McFetridge	Department of Economics, Carleton University, Ottawa
Patrick A. Messerlin	Université de Paris XII et Fondation Nationale des Sciences Politiques (Service d'Etude de l'Activité Economique)
Jean-Louis Mucchielli	Université de Toulon et du Var
Alan M. Rugman	Centre for International Business Studies, Dalhousie University, Halifax
A.E. Safarian	Department of Economics, University of Toronto
Daniel Soulié	Université de Paris — Dauphine
Leonard Waverman	Department of Economics, University of Toronto

Preface

Over the past 15 years much attention has been paid to international technical transfers through large multinational enterprises (MNEs), particularly to transfers between developed and developing countries. Not all such movements of technical knowledge are through MNEs, of course. Important transfers occur, for example, through education of students abroad and through trade in capital goods between unrelated parties. The technology trade in patents, licensing and technical know-how is dominated by MNEs, however. While developing countries are becoming more important as markets for such trade, most of it, like most foreign direct investment, is mainly to and from the OECD countries.

A number of questions about such transfers have been examined in the economic literature. They concern mainly the strategies of MNEs as they seek to optimise the rents on their knowledge, and the strategies of countries as they seek to optimise their national or group welfare objectives. Two of the more important questions are the following.

First, to what extent and under what circumstances are transfers internal, that is, with affiliated firms as distinct from more arm's-length transactions? The emphasis on the internalisation theory of MNEs suggested that firms would prefer those markets where their bargaining power was strongest and those partners with which non-aggressive types of policies could be worked out. This approach has made a major contribution to the theory of MNEs, but it has not told us enough about strategies involving contacts with external partners.

Second, how successful are technical transfers from the viewpoint of the firm and of the country or countries involved? This is partly a question of the industrial structure and capabilities of the receiving country; for example, of the availability of skilled labour and domestic entrepreneurial groups. It is also a reflection of how successfully governments go about their objectives of securing technology on favourable economic terms, or of securing transfer through non-controlling forms of organisation in the face of some preference by many MNEs for internal transfers. In a world of increased competition among MNEs and also among governments for MNE transfers, improved knowledge of MNE strategies is a key to more effective national policies.

This volume is an attempt to add to the knowledge necessary to build a more explicit and more detailed theory of technology transfer. It is based on the papers presented to a conference on 'International Technical Transfers, Multinational Enterprises and National Policies in Advanced Countries' which took place at the University of Paris, Dauphine, from 5 to 7 September, 1985. The conference was sponsored by l'Association de Recherche Economique en Propriété Intellectuelle et Transferts Techniques and the Canadian Economics Association, with the editors of the volume acting on behalf of the organisations involved. We acknowledge gratefully the support of AREPIT, which bore most of the task of organisation, all local arrangements and expenses, and part of the editorial costs. Support is gratefully acknowledged also from the Social Sciences and Humanities Research Council of Canada, which made a grant to cover the travel costs of the Canadian participants, and from the Donner Canadian Foundation for secretarial assistance.

The Conference was the third of a series held by scholars from Canada and France to discuss research of mutual interest. Apart from those presenting papers, a number of researchers from several European countries participated in the discussions. The papers were revised and, where necessary, translated by AREPIT in the period up to about May 1986. Most of the studies involve empirical work testing, among others, the questions on technology transfer posed above. While several studies deal with the experience of particular countries, notably France and Canada, they do so in ways which we believe are of general interest. We offer the volume, accordingly, as a contribution to the international literature on technology transfer.

A.E.S.
G.Y.B.

1

An Interpretive Summary

A.E. Safarian

This comment has two objectives. One is to summarise each contribution to this volume. The other is to highlight some of the implications for national policies on technology transfer. The interpretations on the latter point are my responsibility alone.

The first section of the volume contains three papers which bring out in theoretical and empirical terms some aspects of the relation between technology transfer, multinational enterprises (MNEs) and economic welfare. The second section deals in more detail with MNE strategies on such transfers, and also with a number of related issues of national policy. These findings are explored further in the third section which concentrates on industry studies for particular countries.

TECHNOLOGY TRANSFER AND THE MNE

Jean-Louis Mucchielli uses the concepts of the comparative advantage of the country and the competitive advantage of the firm to integrate analysis of the various forms of international exchange. He begins by noting the comparability of the determinants of the various types of exchange. The differences in supply, demand and market structure which determine trade between countries are utilised to examine the determinants of the multinationalisation of production. This linking of determinants suggests that the international competitiveness of a product is all the more effective when the advantage of firm, sector and country are mutually reinforcing.

Mucchielli then goes on to note, however, that differences in the competitive advantage of the firm and the comparative advantage of the country can lead to relocation abroad. He traces in some detail, with examples, how differences in technology and in factor endowments determine foreign direct investment. While much of the task of testing lies ahead, there are a number of quite specific implications for policy on such investment; for example, in the emphasis on the extent to which, and the circumstances in which, the various forms of international exchange are substitutable.

The paper by Bernadette Madeuf presents and analyses data on the production, and especially the international diffusion, of technology for the OECD countries, including a balance of payments on technology account by country. It then develops indices of the dependence on and competitiveness in technological development, and the trends in comparative advantage in technology for such countries. It is clear that one needs to look at a variety of measures of competitiveness, not just those on technological development. In particular, it is the ability to capitalise on technology, not its production or borrowing *per se*, which is the key relation to economic growth. Madeuf's paper shows, for example, that Japan has used such technology more effectively than the EEC in terms of industrial and commercial development.

The paper by D.C. MacCharles takes this further by underlining how important it is for effective technology transfer and competitiveness that a country have an appropriate framework of policies. One of the themes which recurs constantly is the ways in which foreign trade, as well as direct investment and licences, can serve as both a spur and a vehicle for the transmission of knowledge. Conversely, the benefits of such transfers can be dissipated in an inefficient structure of industry. This is illustrated by reference to the effects of extended protection for Canadian manufacturing, with particularly harmful effects for the productivity of the smaller Canadian-owned firms. Increased competitive pressures in the 1970s and later have led to increased trade specialisation and have largely closed the gap in productivity between Canadian-controlled and foreign-controlled firms. MacCharles devotes the latter part of his paper to the obstacles, notably in organisational terms, of moving still closer to international productivity standards.

MNE STRATEGIES AND NATIONAL POLICIES

An important question in assessing the effectiveness of policies on technology transfer and MNEs is the extent to which such policies take into account the probable reactions of MNEs. Most governments, particularly in the more developed countries, are not attempting to keep out MNEs, except for certain sectors. They are more likely to try to attract them and/or to influence their operations so as to achieve more fully a government's economic and political ends. They are also likely to try to get technology on favourable terms, and, in some countries or sectors, to secure it through non-controlling organisational forms such as licences or joint ventures.[1] Policies designed without taking account of firm strategies are less likely to achieve such ends, or at greater costs.

Three explanations of MNE behaviour have been emphasised in the economic literature. One, the transactions cost approach, considers the MNE as an (at times imperfect) response to imperfect markets for technology transfer. A related strand of theory concentrates on global business strategies. A third strand concentrates on the location decisions of MNEs. The papers noted above and below have much to say on the first two explanations.[2]

The paper by Bernard Bonin presents many of the empirical tests which bring out the differences in technology transfer by various forms of contract. He then compares these other modes of transfer with that within MNEs, from the perspectives of both the buyer and the seller, and taking account of social as well as private effects. Buyers and host governments often prefer non-controlling transfers of technology. MNEs prefer to internalise transfers where that lowers the real costs of technology transfer or increases the rents realised on it. Bonin explores in some detail the evidence on the circumstances in which the more arm's-length types of technology transfer are, or can become, effective substitutes for transfers within MNEs. This central policy question is also picked up in many of the papers which follow.

Gilles Y. Bertin uses a sample of 112 transfers of technology, based on interviews and questionnaires, to examine both inward and outward transfers by MNEs, as well as transfers with affiliates and non-affiliates. The emphasis is on a global MNE strategy regarding such transfers, rather than, as in many other studies, one or other aspect of such a strategy. By looking at different types of potential partners and several objectives, eight types of MNE strategies are examined. These strategies are considered also by country or region of origin of MNEs and six industrial sectors. Among other things, the study concludes that internal partners are preferred to external, and that, among the latter, smaller domestic firms and those outside the 'core' activities of the transferring firm are preferred. These conclusions, along with those in several other papers in this volume, suggest that the transactions cost approach to technology transfer is still of considerable interest, despite some evidence that 'new' (i.e. arm's-length) types of technology transfer have increased in recent years.[3] Governments which attempt to separate technology from effective control by MNEs will continue to have problems where newer or more radical technologies are involved (see McFetridge below), or, as Bertin suggests, in core technologies or where the recipient is a potentially strong competitor. Bertin suggests also that these findings carry important nuances: for example, firms are more likely to sell to local partners who offer information in return, and it is easier to get such exchanges in truly globally competing industries.

One of the strategic decisions for an MNE is the level and location of its inventive activities. Such activities are much less decentralised in MNEs than are production and employment. Many governments, in turn, have displayed an active interest in attracting or retaining such activities. The main focus of the paper by Hamid Etemad and Louise Séguin Dulude is the pattern of decentralisation of inventive activity in MNEs, measured in terms of patenting. The sample data for the tests was drawn from the data bank of Consumer and Corporate Affairs Canada. Determinants of decentralisation which are considered include the MNE's overall inventive activity, the extent of its foreign production, the main sector of activity and the MNE's region of origin. The combined effect of the first two determinants receives more support in these tests than do the other variables.

Alan M. Rugman shows how the traditional patterns of FDI have been reversed in Canada over the past decade, with outflows now well in excess of inflows. Both sets of flows are largely with the USA, a matter of some political concern on the Canadian side. He regards the substantial outflows from Canada in recent years mainly as part of the general upsurge of FDI in the US market. The latter, in turn, reflects cross-investment by MNEs in each other's home markets. Turning to micro analysis, Rugman examines how 16 of the largest Canadian MNEs have succeeded in developing and protecting firm-specific advantages which allow them to compete globally. For almost all of them, the firm-specific advantages are largely based on marketing and management skills rather than technological intensity. This in turn reflects country-specific advantages, notably the natural resource base. Rugman is accordingly critical of the heavy emphasis on technological development as such which dominates much policy discussion in Canada regarding international competitiveness.

The paper by D.G. McFetridge examines empirically a number of propositions on technology transfer, both by using the existing tests and introducing some further ones utilising the Harvard Multinational Enterprises Database. The research was motivated by a number of concerns of Canadian economic policy but the findings are quite general. His conclusions are that transfer lags have shortened over the past two decades and relatively more arm's-length transfers occur now. The timing and the mode of transfer are jointly determined, however.

Initial tests suggest that national economic, technical and policy determinants also affect the timing and mode of transfer. McFetridge examines these determinants in some detail, with a number of interesting policy implications. For example, countries which do not have screening or equity controls on FDI tend to be earlier in the transfer order than those with such controls, although present tests suggest there is no difference between those with some controls and those with controls which are pervasive.[4] McFetridge notes also that the apparent compression of the product cycle reduces the potential pay-off from targeting high-tech strategies. In the case of Canada he notes that domestic manufacture of new products has been introduced as quickly as in other developed countries, while the technology of manufacturing has adapted relatively slowly.

INDUSTRY STUDIES

One's attitude to the welfare effects of MNEs depends in part on whether they are considered to limit competition in products and technology markets, or whether they compensate for imperfections in arm's-length transfers hence extend the possibilities for efficient exchange.[5] A number of the papers noted above deal with aspects of this issue, mostly concluding, with some qualifications, in favour of the latter interpretation.

The paper by H.C. Eastman examines this and related welfare issues with respect to the impact of the international pharmaceutical industry on a country such as Canada, which is a major host country for such MNEs. He notes that there are two obstacles to social efficiency in R&D: duplication of research can lead to excessive investment, while appropriability of the return on such investment is imperfect where imitation occurs. Patent protection is intended to deal with these problems. It also delays the appearance of competitive products or encourages product differentiation with little therapeutic improvement.

Governments in almost all countries attempt to modify the resulting problems by such devices as control of drug prices or profits, limits on reimbursement in drug insurance schemes, or weakening patent protection. The device used in Canada is compulsory licensing of patented drugs for a fixed royalty. This has stimulated competition by generic firms, leading to significant savings to Canadian consumers and governments. Eastman notes that the fixed royalty does not cover the cost of research abroad which is incurred by MNEs operating in Canada. He suggests that the balance between assuring an adequate return to invention and extending the benefits of technical progress can be restored, and other problems avoided, by retaining compulsory licensing with an increased fixed royalty rate. The issue which is highlighted in this paper is that of policy design intended to reconcile the interests of national (consumer and taxpayer) groups with producer and international welfare.

Patrick A. Messerlin examines the benefits and costs of a policy of import protection combined with foreign investment regulation as a method of assuring technology transfer by MNEs. The case examined is Japanese direct investment in France, particularly in the electronics industry which is subject to a variety of such measures. While the policy attracted foreign direct investment, it is not yet clear whether it has succeeded in avoiding the pitfalls of such an approach — in particular, the economic costs of protection evident from the paper by MacCharles on Canada. Messerlin measures some of the effects of such a policy approach and indicates what still needs to be done to complete such measures.

France is not Canada, and, in particular, has a unitary government and far greater financial and industrial power centred in that government. On the other hand, France is not Japan, either, a country which has succeeded in developing some efficient global industries by policies which may prove difficult to export to other countries.[6] Thus Daniel Soulié uses data on foreign trade, patents and technology payments to assess the adaptation of the French automobile industry to technical change. He examines the extent to which the industry's position has deteriorated significantly internationally in terms of technical change, whether one thinks of products and processes, on the one hand, or management and strategic behaviour, on the other. This relative deterioration he ascribes to a failure to realise fully the opportunities and constraints arising from globalisation of the industry, including its parts sector.

It is a failure he relates in turn to the conflicting relationship between automobile producers and producers of automotive equipment.

The paper by Melvyn Fuss and Leonard Waverman examines the joint venture as a means of transferring technology, with reference particularly to Japanese joint ventures in automobile production in North America. They challenge existing studies of the comparative advantage of Japanese automobile producers, reminding us of the crucial distinction between differences in production costs and differences in productivity. Their detailed estimates of possible sources of the Japanese cost advantage suggest most of it is in the difference in input costs and in changes in the yen/dollar rate rather than in productivity differences as such. The productivity difference which would be transferred to the USA by Japanese direct investment is a relatively small part of the overall advantage enjoyed by Japan. Fuss and Waverman question, therefore, whether policies encouraging such direct investment will do much to raise welfare, especially if there are adverse competitive effects from increased concentration in national auto markets. Another implication of their study, however, is that Japanese joint ventures might be able to write labour contracts in ways not open to existing US producers. Thus technology transfer by MNEs may change relative factor prices by ways other than through evening the international differences in factor supplies or productivity.

A number of the authors emphasise that further empirical work is under way; hence their results are still tentative. Nevertheless, overall one can suggest that a combined transactions cost–business strategy approach proves a useful analytical approach. There is a confirmation of some of the earlier work on internalisation theory, but also a recognition of the limits to such an approach given the complex strategic decisions faced by firms in an increasingly competitive global market. There is more emphasis in the volume on MNE strategies than government strategies. Nevertheless, the evidence in several studies underlines the view that governments face considerable difficulty in transferring (or duplicating) many of the newer or more complex technologies outside the MNE organisational form, especially if a core technology is involved or the recipient firm is a potentially strong competitor. That said, there is still scope for affecting the terms of exchange, especially where competition exists or effective intervention is possible; non-controlling forms of transfer play a growing role in the strategy of many types of firms; and appropriate complementary industrial policies are critical to efficient transfer.

With significant exceptions, the emphasis in the studies is more on MNE strategies and some related national policy issues, rather than the design of optimum and feasible policy instruments. The latter would have required more focus on the determinants of public policy, on personal and regional income distribution issues, and on home-country effects and the reactions to host-country policies. While much remains to be done, the volume presents a considerable harvest of ideas and evidence which should be of interest both to theory and policy on technology transfer issues.

NOTES

1. For studies on government policies towards MNEs, see Bonin (1984), Guisinger (1985), and Safarian (1983).

2. Rugman (1981) and Caves (1982) are recent contributions to the transactions cost approach, which owes much to Williamson (1975). Porter (1980) is among those who have contributed to competitive strategies of MNEs, involving theories which are also much analysed in the literature on domestic firms. The location decisions of MNEs are now given considerable emphasis by international trade economists. See also Mucchielli, Chapter 2 of this volume, and Dunning's (1979) attempt at an integrated approach.

3. See Chapter 9 by McFetridge is this volume, and also Oman (1984).

4. It would be interesting to know if policy restrictions net of incentives, as proposed by Guisinger (1985), would change a country's transfer order.

5. This contrast in views is clearly brought out in Dunning and Rugman (1985).

6. The paper by Fuss and Waverman, Chapter 13 of this volume, adds a further dimension to this point.

Part One: Technology Transfers and the Multinational Enterprise

2

Multinational Enterprises, International Investments and Transfers of Technology: the Elements of an Integrated Approach[1]

Jean-Louis Mucchielli

The controversies between the authors who put forward the initial proposals for an integrated summary of the debates on trade and investment have centred on the incorporation of comparative advantage in the analysis of international investments. It is this element which we want to build into our integrated analysis. This is not a question of finding a halfway house between the approaches of Dunning (1981) and Kojima (1978). It seems necessary to go beyond these analyses and propose a new combination of the various types of advantage, built around the notions of the comparative advantage of the country and the competitive advantage of the firm.

As a first stage, what we need to do is to relate to each other, and go a little more deeply into, the common factors governing the international exchanges of both goods and investment. The next stage will be to look more closely at the differences between them.

THE FOUNDATIONS OF THE INTEGRATED ANALYSIS

Inter-country differences as the basis for international flows

The main agent involved in exchanges of goods, as of investments, is the firm. It is the firm which exports and imports, invests abroad or receives investment from abroad. These investments, it should be recalled, are 'technological packages'; they consist not only of financial capital but also of technology, know-how and services, and lead to transfers of skilled personnel (engineers, etc.) and of machines, in other words capital goods. Moreover, in this latter case, the frontier between exports of merchandise and of capital equipment becomes somewhat hazy. Even the breakdown between consumption and production goods becomes inadequate to capture the distinction.

These few elements are sufficient to indicate that the decisive factors in exchanges of goods and of investments are comparable. The analysis of the factors involved in merchandise trade should, therefore, help us to define and understand those which generate foreign investment.

If countries, groups or individuals carry out exchanges, this must be because they differ from each other. This is reflected in a divergence in the relative prices of the products and it is this divergence which leads to the exchanges. It is precisely because a good can be obtained elsewhere at a lower relative price that trade with the partner takes place. These divergences (explicit or implicit) in relative prices are in fact the reflection of more deep-seated differences, based on the economic characteristics of potential co-traders. This is the basis for the notion of comparative advantage in trade and for specialisation.

Applying the converse reasoning, this time at the level of the national economies, it can be stated that when the following six conditions are simultaneously satisfied, there will be no exchanges, because there will be no differences in the relative prices of products:

(a) production functions are similar as between countries;
(b) factor endowments are identical as between trading partners;
(c) tastes of economic agents in different countries are similar;
(d) there are no possibilities of economies of scale in the production of all goods;
(e) markets for finished products are not distorted;
(f) markets for the factors of production are not distorted.

If, on the contrary, one of these conditions is not met — or *a fortiori* several of them — differences in relative prices appear and exchanges take place, since each partner can benefit from comparative advantages in one or more products *vis-à-vis* the other co-traders.

Traditionally, the simplification needed in the construction of any theoretical model, attempting to grasp part of a complex reality by using simplified concepts, has meant that each approach has confined itself to the analysis of just one of these differences. The Ricardian model, for example, examines the trade resulting from differences in technology, in which comparative advantage is based on technical knowledge in the hands of only one partner to the exchange. In Ricardo's famous illustration, Portugal is more capable of making both one unit of cloth and one unit of wine, even if the countries are identical in all other respects (human resources, tastes, etc.). The possible disadvantage of this approach resides in the fact that it takes the difference in technology as given; the basic model being static, it could hardly be otherwise. But with neo-technological refinements demonstrating the importance of innovation as a means of acquiring comparative advantage, we find ourselves with the analyses of Posner (1971) and Vernon (1966, 1979).

The Heckscher-Ohlin model, for its part, is based on differences in factor endowments. Here again the initial differences are given, but they are themselves determined by differential growth in the factors of production, through the quantitative evolution in the stock of machine capital or human capital. The whole of the neo-factorial analysis rests on these changes.

Differences in tastes between agents can also give rise to exchange. In this case everything would be identical between the co-traders except the respective demands for each good. Ohlin (1968, pp. 65–6) had already studied this phenomenon, taking as his example Danish imports of butter from Siberia and Danish exports of home-produced butter to Great Britain, 'by reason of differences in taste'. In this particular case, the exchanges can even be within a sector and within a product type. The various elements involved in this type of trade can be combined through the concept of 'demand difference' put forward by B. Lassudrie-Duchene (1971).[2] It should be stressed that in this approach the relative levels of consumption for each good are no longer the same. Since relative demand is different, relative prices will diverge, even though endowment, techniques and technologies are all the same. In fact there is no longer congruence between domestic and foreign demand.

Differences in relative prices are also created by the existence of economies of scale. Following the initial introduction into the analysis by Graham (1925), this notion was at first used to assess the distribution between countries of gains from trade, some countries becoming specialised in goods produced with decreasing returns to scale and others in goods with increasing returns. The losses from trade resulting from specialisation of the former were then put forward as a justification for protection.

With identical technology, similar relative endowments, similar tastes, etc., a large country has a comparative advantage for a product where there is a 'threshold' in the cost curve. The large domestic market allows it to be more competitive than other countries for these products; average costs are lower and so are relative prices. The USA is perfectly capable of taking advantage of this difference in the aeronautics industry, *ceteris paribus.*[3]

Moreover, it is known that, in automobiles, increasing returns to scale appear discontinuously in quantitative leaps and in very different fashion for assembly operations and the manufacture of major parts. A calculation in the case of the British industry shows manufacturing cost falling by 15 to 20 per cent when production goes from 50,000 to 100,000 vehicles per year, by 5 to 6 per cent when it goes from 125,000 to 175,000 vehicles, and by 17 per cent when it goes from 300,000 to 1,200,000 units per year. More generally, the calculations show that the minimum size for efficiency in assembly is 250,000 units per year and in manufacture two million vehicles per year (United Nations, 1983, p. 20). In the hypothetical case of autarky, any country with domestic demand below these figures would be at a disadvantage in terms of production costs compared with a large country. In conditions of full international trade, all means will be used to reach the optimal (or maximum) production level from the point of view of economies of scale. The 'world car' is just such an attempt.

From the point of view of gains from trade, the small country, contrary to the Heckscher-Ohlin model, is in danger of losing out. It will then be in its interests to specialise in standardised products for which world demand is strong and homogeneous. An example is the observations applied by Dreze (1960) in the Belgian case.

The two remaining types of differences involve imperfections in the markets for finished products and for factors of production. The analysis of imperfections in the market for finished products takes us back to the theory of oligopoly but also to that of monopolistic competition. In the latter case, products are no longer homogeneous, being differentiated notably by their quality. The difference no longer resides in the production of the product, but in the characteristics of consumption in the sense used by Lancaster (1980), exploited in a particular market structure.[4] The two types of distortion, non-homogeneity of products and imperfect competition, may be combined; the latter can exist without the former, but not the reverse. The concept of 'demand difference' can of course be reintroduced in this new market structure, in the form of 'demand differentiation', recognising that it is the market structure which in this case creates the divergence in relative prices.

Finally, allowance must be made for distortions of factor markets, which alter their respective returns compared with a system of pure and perfect competition. These distortions may be exogenous in origin; for example, greater state intervention in one country than another, monopolies of factor supply such as a trade union, and the existence of different wages in the towns and in the countryside for the same quality of work.[5] Each of these distortions generates differences in the relative factor returns as between sectors and countries. They also lead to a delinking of these returns from the relative physical factor endowments. The country may then be induced to specialise in products for which it has no real comparative advantage. It may also use the wrong production techniques, by comparison with an economically efficient situation (Krueger, 1977). The distortions may be endogenous to the economic variables themselves. In that case, it is preferable to talk of differential remuneration between sectors.[6] This occurs when a factor is immobile in a given sector in relation to others. As we have seen, this means factor specificity, a situation which generates a differential in relative remuneration between sectors and between countries as a function of each partner's relative endowment in the factor and of the elasticities of substitution between the mobile factors and the specific factors. All these elements in the factor markets in turn affect the relative prices of the products and create part of the differences underlying the exchanges.

To our knowledge, only one author has attempted to explain international trade in terms of the resemblance between countries and not the differences. Linder (1961) aims to explain exchanges between industrialised countries by means of 'representative domestic demand'. Countries with similar incomes then have relatively similar representative demand, 'the range of exportable goods is identical to or included in the range of importable goods' (p. 91). When Linder looks at the cost differences which may exist between so-called similar countries, however, he attributes them to 'advantage in the possibilities for exploiting raw materials, entrepreneurial skills, economies of scale . . . factor endowments' (pp. 103–4). Quality differences, for their part, are generated mainly through monopolistic competition. We are back again to the elements just discussed. Furthermore, by rejecting traditional trade analysis, Linder in the end finds himself examining the volume of exchange and no longer its nature, as pointed out by Bhagwati (1964, p. 28). The structure of exchange within importable and exportable goods is then, for Linder, the result of some 'historic accident' and is likely to be 'highly volatile' (p. 104).

It should also be remembered that domestic demand is by no means overlooked in the basic models. Otherwise, how would the quantities traded, the prices and the different specialisations be arrived at? Demand is even representative in Linder's sense, since production initially takes place in conditions of autarky on the basis of this domestic demand. Later, the demands of the different countries are met, since, having the same tastes, they exchange goods which each of them had been producing under autarky.

Exchanges of products and of international investment: similarity of the determining factors

These major groups of determinants noted above recur in the explanations to be found for the multinationalisation of firms. All that is needed is to examine each of the six differences between countries which underly trade from the point of view of the reasons for industrial relocation.

Differences in technology

This difference between country of origin of the firm and the host country lies behind two explanations of relocation. In the first case, establishment abroad is the consecration of a technological advance by the parent company over firms in the same sector in the host country. This is simply an aspect of the product cycle. The choice between exporting, selling licences or relocating production is a function of factors associated with the desire to maintain this technological advantage as long as possible. In the second case, the aim is the acquisition of a new technology. The firm then establishes itself in a country where the technology in the sector concerned is more advanced than in the country of origin, with a view to acquiring these techniques in a learning-by-doing process. The second thesis has been put forward especially by Franko (1971) who estimates that at least 50 per cent of the industries setting up in the USA expect in this way to acquire new technological or organisational skills. The same motivations are to be seen in certain cases of US investments in other developed countries and of intra-European movements.

Another phenomenon, connected with 'forgetting-by-not-doing', can also come into the picture and induce, within a given sector, relocation to a country with a different technology. As early as 1971, Bhagwati (p. 457) envisaged the possibility of cross-investments between Ford and Toyota, the former trying to acquire the most efficient techniques for the manufacture of small cars and the latter attempting to become familiar with the construction of large ones. The agreements signed between Chrysler and Mitsubishi at the beginning of the 1970s on reciprocal marketing of cars also involved this technological aspect.

From the point of view of the types of direct investment abroad — for example, creation of subsidiaries, buyout of foreign firms, or acquisition of shareholdings — a firm enjoying a technological advantage will more readily think of setting up its own subsidiaries, whereas in the opposite case, the investment will normally take the form of total or partial acquisition of foreign firms. Behaviour will also be influenced by the nature of the sector and the country concerned.

Differences in factor endowments

In this case a firm relocates in a country where relatively abundant factor endowment gives the possibility of lower costs. These investments are generally linked to the existence of 'cheap labour'. Labour costs differ substantially between sectors even in the developed countries, so that relocation does not necessarily take place from the country which is the more developed overall to the less developed one. Similarly, relative factor abundance can apply to skilled manpower which is less mobile internationally than other capital. In this case, the relocation will be aimed at benefiting from the advantage in skilled labour.[7] Sometimes these two types of factor abundance are exploited at the same time. For example, Ford in West Germany employs 18,000 immigrant workers out of a total workforce of 20,000 (low-cost manpower), while at the same time benefiting from the efficiency and the technical and other supervision provided by the skilled German workers.[8] Finally relocation can be associated with the exploitation of certain absolute advantages, for example, the extraction of raw materials.

Differences in tastes

Demand intervenes in various ways in the relocation process, but is treated here simply from the point of view of differences in tastes. One of the motives for setting up abroad is the need to be closer to local demand so as to have better knowledge of it, and if necessary, to adjust part of the firm's production to the characteristics of that demand. The investment might initially take the form of the establishment of commercial subsidaries, followed later by production. Conversely, by coming closer to customers, the firm may contribute to 'cultivating a taste for difference', by exploiting this type of demand at closer range. The leading French fashion houses, for example, will make play with 'la mode parisienne' in their US production and sales subsidiaries.[9]

Economies of scale

This factor seems to be important, although controversial (for example, Hirsch, 1976). Economies of scale can be linked with the size of the host-country market; this played a major role in the relocation of US firms to the EEC, just as it now does in the relocation of European firms to the USA (Scaperlanda and Mauer, 1969).[10] The larger the host-country market, or the faster its expansion, the lower the fixed costs of establishment in relation to expected sales and profits. Relocation then takes place all the more easily in that the size of the host-country market makes it cheaper to produce locally than to export. However, the manufacture of products or sub-assemblies which can quickly achieve economies of scale will be easier to relocate because the efficiency threshold will be lower than for other products. This is one explanation for the relocation of vehicle assembly rather than complete manufacture (United Nations, 1983).

The possibilities for multinational enterprises to exploit economies of scale becomes an advantage over the domestic firms in each country since the markets supplied by the former generally cover several countries. Economies of scale are also to be found in the organisational structures of the MNE, which are usually more rationalised than in other firms. These elements lead to analysis of the forms taken by the markets.

Distortions in product markets

These distortions correspond to barriers to pure and perfect competition. While incorporated in trade theory, they have been more developed in the analysis of the MNE, thanks to the contribution of industrial organisation theory.

Distortions in product markets bring out the importance of oligopoly. Relocation decisions appear to be linked to international oligopolies, through behaviour involving imitation, differentiation, defensive strategy or any other approach based on sharing the international market and on the growth of the firm.

Certain economic policies can also contribute to the creation of this type of distortion. This is true, for example, of tariff or other commercial barriers set up by the host country, which the firm will try to get around by setting up subsidiaries. This process helps explain Japanese relocation operations in Europe and the USA at the present time. The same is true of certain US investments in Europe, and vice versa. In the country of origin the attack on distortions, as in the case of US Anti-Trust Laws, sometimes induces firms to relocate simply in order to grow. This phenomenon heightens the specific character of international investment, since the growth of conglomerates in the home country is not affected by the anti-trust arrangements.

Distortions and differentials in remuneration in the factor markets

These distortions can emerge for the three broadly-defined factors of production, labour, capital and technology, as the result of exogenous or endogenous factors. In the case of labour markets, for example, unionisation may induce a firm to relocate or to multiply the number of its production sites in order to reduce the risk of conflict. In capital markets, special bonuses or the package of arrangements available to foreign capital can also encourage relocation. Similarly, the desire to have better access to international capital markets, or to obtain protection against taxes on profits, will create a number of investments near the principal financial markets or in tax havens. Lastly, sector immobility of certain factors, or their specific nature, will lead to differences in their relative remuneration. This applies to virtually all foreign investments, which are specific by nature; the sectors in which they are set up are usually the same as those in the countries they come from.

COMPARATIVE AND COMPETITIVE ADVANTAGES IN INTERNATIONAL INVESTMENTS

Levels of analysis and determining factors in exchange

The three levels of analysis — sectors, firms, countries — are to be found in the determining factors set out above, both for commercial exchanges and for international investments.

Comparative and competitive advantages

It is possible at this stage to discern a certain analytical structuring of these levels. The firm is in possession of competitive advantages due to its intrinsic characteristics, such as a certain quantity of human capital, a system of organisation, an R&D structure, and a number of new products. These can constitute an absolute advantage over the firm's competitors. The market structure is a particular characteristic of the product or sector in question, which will often go hand in hand with the nature of the product. A technology-intensive product or sector will more easily lead to an oligopolistic market structure than will a standard one, for which competition will be intense. Finally, a country has comparative advantages in relation to others, resulting from its supply capacity as well as its demand possibilities as discussed earlier.

The analysis can then take place simultaneously at the level of the firm, the sector (or product) and the country. The economic variables for a national economy are simply the supply, the demand and the market structure. Nevertheless, the firm is going to supply products and demand factors of production, whereas the home or foreign country, as a whole, will demand products (through consumers) and supply factors of production. In this way, the entrepreneur will have to find the best combination of the firm's own resources (demand and supply) with those of the country of origin, in order to remain competitive in the product he manufactures and sells on national and international markets. The international competitiveness of a product will be all the more effective if the competitive advantage of the firm and the comparative advantage of the country are mutually reinforcing.

The first case to be considered is that of technology. Comparative advantage can be associated with the technological advance of a country as a whole over its partners, explained by the frequency of inventions or innovations, supported by major overall expenditure on R&D and an effective system of fundamental and applied research, among other variables. A country's technological advance can also involve a single sector in which innovations have come to be concentrated over time. The German chemical industry is a case in point. At the international level, some sectors may be more apt than others to generate innovations. We are here once more concerned with the industrial cycle and the notion of technology-intensive sectors. Lastly, technological advance can exist for only one or a few firms in the country, those with a higher capacity to innovate not only because of more intense R&D policies but also because of the entrepreneurial propensity to innovate. Innovation is not necessarily confined to the large firms.

The comparative technological advantage of the country is then all the greater and all the more solid if there is synchronisation between the three levels: innovative firms in an innovative sector in a technologically-advanced country create unparalleled export competitiveness. The USA was probably in this ideal situation after the Second World War. The advantage is comparative because it is measured in relation to other countries in all the technology-intensive goods and sectors; technological advance, on the other hand, is an absolute.

Relative advantage and sectoral stratification

Other situations are possible in which competitiveness is less strong. If technological gaps between countries shrink or disappear, with no country occupying the first rank, an economy can only establish comparative advantage in those sectors where firms are especially innovative. For the totality of technological products or sectors, comparative advantage is divided among countries which have reached the same technological level. This leads to cross-trade in relatively technology-intensive goods, within as well as between sectors. If technological advantage applies only to the firm and not to the country or the sector, comparative advantage disappears and the firm is increasingly hard pressed to exploit its own competitive advantage through exports. The same reasoning holds good for the other determining factors in trade, especially those which underlie real comparative advantage, in other words whose which are not created by some exogenous distortion.

For relative differences in factor endowment, comparative advantage is all the more important in the case of a skilled-labour-intensive product. For example, to the extent that the country is itself well-endowed with this factor, the sectors incorporating this type of labour are developed in the country and the many firms well supplied with it apply the appropriate factor intensity.[11] Similarly, if the relative abundance of skilled manpower becomes comparable in all developed countries, the comparative advantages will be reduced to inter- and intra-sectoral differences in this factor. If only the final link in the chain is left, i.e. a skilled-labour-intensive firm operating in a sector and a country without this characteristic, then it will be difficult, if not impossible, for the firm to keep its competitive advantage for any length of time, because of the uniqueness of its endowment.

In the case of taste differences, a country will have more comparative advantage to the extent that its population is large and diverse, so that marked taste differences are already to be seen on its domestic market; to the extent that the sectors involved lend themselves to substantial real or artificial product differentiation; and to the extent that the firms are themselves disposed to extend the range of their products. The same reasoning applies to economies of scale and to the case of large firms or production units in sectors favouring this type of cost structure and in an economy which is itself large.

In these various situations a country's comparative advantage for a sector or a product and the competitive advantage of a firm are always mutually reinforcing if the three levels are in perfect tune. Otherwise the advantages are reduced, and may even disappear entirely. The principle of comparative advantage applies to all countries. Even if a country has no sector in this situation, it can still obtain a specialisation advantage in sectors where the discordance between the three levels is least.

Market imperfections can be treated as a factor in their own right, as a determining factor in trade, or as elements superimposing themselves on the earlier ones. This is the case for oligopolistic and monopolistic structures, the latter occurring often in the case of technological advantage. As a pure — i.e. sufficient — determining factor, one can quote the case of the natural monopolies associated with ownership of the unique source of a raw material.

For all these types of determining factor, the same three characteristics (supply, demand, market structure) and the same three levels (country, sector, firm) apply. Firms and countries will be in phase for sectors in which the former have a competitive advantage and the latter a comparative advantage, the sector or product group in question being then fully competitive at the international level.

Conditions for foreign investment and integrated analysis: two examples

International investments and technological differences

Up to now, the factors determining exports and investments have appeared as entirely similar. But this fails to explain why a country can be specialised in exports rather than in producing abroad, or why a firm will be an exporter of a given product rather than produce it abroad. From now on the firm, instead of acting only at the level of international exchanges, is assumed also to look ahead in arranging its production activities. In the first situation, product demand and market structure are already international; now, factor supply becomes international as well. The international choice of production site has to take this new factor into account. While the advantages of commercial specialisation are greatest when the competitive and comparative advantages are in phase with each other, the logic of the new situation is the reverse: it is the discrepancy between the two types of advantage which creates the conditions for relocation. The most frequent cases are when the firm's demand for factors cannot be met by supply within the country of origin, or when supply by the firm cannot be sufficiently absorbed by the country's demand for the goods. Market structures will have all the greater influence on the relocation aspects, the greater is the difference between the domestic and world markets.

This analysis can be illustrated by looking at the principal determining factors in location in the case of differences in technology, as shown in Table 2.1. In the first situation shown in the table, the firm, which wants to become or to stay competitive on the international markets for high-technology products, has a strong demand for this factor which cannot be met locally because its country of origin has not the necessary supply. At the level of country of origin there is a discrepancy between the characteristics of the firm and the sector, on the one hand, and those of the country, on the other. The firm relocates to a country which will supply the technology it needs. There is upward relocation to the technologically more-advanced country, in order to acquire the technology through learning-by-doing. This is what we have been seeing in the case of several foreign investments in the USA.

The demand element is reintroduced if one takes, for example, the case of a firm wanting to exploit fast-growing foreign demand. It will then be better placed to meet part of its own domestic demand when this emerges. The firm will therefore have stolen a march on its national rivals.

The establishment of South Korean electronics firms in the USA is a perfect example of this. Samsung, the leading Korean firm in this sector, wanted to move from small-scale electronics to the more sophisticated manufacture of computer memories. It therefore set up in Silicon Valley in July 1983, in order to try to 'reduce the technological gap with the Americans and Japanese from ten years to three', according to its managers (*Economist*, 1984).

Table 2.1: Determining factors in relocation in the case of differences in technology

Firm	Sector/Product			Country of Origin		Host Country	
Competitive Advantage	Domestic Market	Overseas Market	World Market	Comparative Advantage		Comparative Advantage	
Product Supply Technology Demand	Market Structure (Supply and Nature of Products)			Product Demand	Technology Supply	Product Demand	Technology Supply
1. +	–	+	+	–	–	+	+
Potential technological advantages: strong demand for technology	Technological industries International oligopolistic competition Frequent new products Domestic market still weak			Potential unrevealed demand	Technology supply non-existent	Substantial product demand High income	Substantial technology supply Innovative country
	'Learning by delocating', location in higher-technology countries for the sector concerned						
2. +	–	+	+	–	+	+	–
Effective technological advantage		Idem (1)		Domestic demand still non-existent	Substantial domestic technology supply	Strong demand for product High incomes	Technology supply still non-existent
	Japanese product cycle; location in country with higher demand						

3. + Idem (2)	– Product already standard	+ Product still new	+ Keener competition	– Domestic demand increasing	+ Idem (2)	+ Product demand increasing	– Technology supply still weak
	Delocation phase of the normal product cycle in the lower-technology country						
4. + Advantage in appropriate technology	– Product and process standard or no longer existing on international markets	+	–	– Domestic demand decreasing	+ Domestic supply of appropriate technology	+ Product demand increasing	– Technology supply weak
	Relocation towards developing countries with the appropriate technology						

The second case examined in the table is that of the 'Japanese product cycle' developed by Tsurumi (1973). According to this theory, because of the lag in domestic demand some of the more sophisticated Japanese products are exploited in the USA even before they are sold in Japan. The author quotes the examples of colour television and 'mini' hi-fi electronic equipment. In this case there is a distortion between the technological capacities of the firm and the country of origin, and the domestic market. The supply of products by the firm no longer corresponds with the country's demand for the products. Abroad, the firm finds itself in the reverse situation: the demand is there, but the local supply of the technology factor, as of the product itself, is still non-existent. This out-of-step situation could be solved by exporting all the Japanese production to the USA. However, this recalls the situation described in the first phase of Vernon's (1966) product cycle: the firm needs to be near its market and its demand in order to adjust and modify the characteristics of the product being supplied. The Japanese producers therefore relocate production so as to be near the demand, while still importing from Japan the most technology-intensive components of the product.

The third case corresponds to the relocation phase of the classic product cycle. There is now a discrepancy between the supply of the product by the firm and national demand for the product. This leads first to export and then relocation to a country where demand for the product is growing, but where as yet the supply of technology available nationally is not such as to facilitate competitive production from local firms. The relocation is then downward with respect to this technology and in this sector. This fits the case of certain European investments in the USA, such as the installation of Lafarge-Coppée through the buyout of General Portland (Nalin and Defava, 1984). This French firm is technologically far in advance of the US firms, making cement by the so-called dry process, which permits substantial economies of energy and much less pollution by comparison with the normal wet process. Moreover, the Florida sun-belt site for the implantation means that it is located in a rapidly-expanding market, in contrast to that of Europe. Lastly, relocation was also encouraged by the high costs of transporting cement, which mean that production must take place as close as possible to the final consumer.

The last case examined is that of a relocation of appropriate technology. The firm here has a competitive advantage in this technology although domestic demand for it, or for products based on it, is declining, while remaining strong in countries which are not as well developed. Relocation then brings the various levels of analysis back into phase, especially as between the firm's advantages and those of the country of relocation for the given sector or product.

For each of the cases analysed, therefore, foreign investment results in maximising the positive (+) situations for all the characteristics shown in the table and minimising the negative (−) ones.

International investments and differences in factor endowments

In examining differences in factor endowments, the same relocation logic applies as in the previous case. There will be a discrepancy between the characteristics of the firm and those of the country of origin and this discordance can again be intensified or reduced by the characteristics of the markets and the nature of the products. These elements apply to all types of factor endowments, including skilled and unskilled labour, capital, or raw materials. Table 2.2 summarises most of the possible cases.

Concerning endowment in unskilled labour, two situations are shown: both assume that the firm produces an unskilled-labour-intensive good and both bring out a distortion between the firm's needs for this type of labour, if it wants to remain competitive, and the shortage of national supply. Relocation then takes place towards a country where unskilled labour is plentiful and not expensive. In the first case, however, this distortion is accompanied by a second discrepancy concerning the demand for the product. In the national economy this demand is weak, whereas it is strong in the host country. By relocating, the firm takes advantage both of the lower costs and of the strong local demand which provides it with immediate outlets. In the second case, on the other hand, the firm relocates uniquely in order to match its demand for labour with the local supply. The manufactured product is then re-exported to the national or international market, since there is no demand in the host country. This situation is less effective from the firm's point of view if international demand for a standard product is itself on the decrease, making the outlets much less certain.

These two possibilities recall the behaviour of French and Japanese firms. The former tend to relocate to take advantage of cheap labour in the French sphere of influence in Africa, while the latter relocate in South-East Asia and Latin America. The French MNEs, although they reduce their costs, benefit only to a small extent from local sales of such products as textiles and small-scale

Table 2.2: Differences in factor endowment, discrepancies, competitive and comparative advantages and relocation

	Firms	Sectors/Products			Country of Origin		Host Country	
	Competitive advantage	Domestic market	Overseas market	International market	Comparative advantages		Comparative advantages	
	Demand for factors or domestic availability of factors (specific factors)	Market structure Supply of products			Product demand	Factor supply	Product demand	Factor supply
	1. +	–	+	+	–	–	+	+
Unskilled labour endowment	Specific endowments in unskilled labour	Industry declining Standard product Unskilled-labour-intensive techniques Keen international price-based competition Active international markets Domestic market decreasing			Decreasing product demand	Shortage of cheap unskilled labour	Increasing demand	Abundant unskilled labour
							Relocation to regain cost and demand advantages	
	2. +	–	+	+	–	–	–	+
	Specific endowments in unskilled labour	Idem (1)			Decreasing product demand	Idem (1)	Demand for product non-existent	Abundant unskilled labour
							Relocation to regain cost advantages: re-export; free trade zones	
	3. +	–	+	+	+	–	–	+
Skilled labour endowment	Specific endowments in skilled labour	Idem (1)			Strong demand for product	Skilled labour rare and/or costly	Demand not yet revealed	Abundant skilled labour
							Relocation to countries with abundant skilled labour	

Capital endowment	4. + Large factor demand for capital and machinery	+	+ Mature industry Capital-intensive products Oligopolistic competition	+	− Demand weak	− Poorly-endowed in capital	+ Strong demand for intensive products	+ Well-endowed in capital
							Upward relocation to obtain capital	
	5. + Substantial endowment in specific capital	+	+ Idem (4)	+	+ Substantial demand for product	+ Capital available	+ Strong demand	− Poorly-endowed in capital
							Downward relocation to exploit specific endowment in capital	
Endowment in raw materials for intermediate goods	6. + Specific endowments in skilled labour, capital, the technology necessary for export of raw materials	+	− Basic industry Monopoly, cartel Technology- or capital-intensive techniques Standard undifferentiable intermediate products (oil, copper, lead, etc.)	+	+ Strong demand for products	− Supply non-existent Non-availability	− Demand non-existent	+ Raw material deposits
							Relocation to exploit the deposits. Initial processing stages possible locally (chemicals, refineries)	
Endowment in raw materials for finished products	7. + Idem (6)	+	− Processing industries Monopoly, cartel, oligopolies Capital-intensive techniques Non-technological products, but highly differentiable at final stage (foodstuffs, drinks, etc.)	+	+ Strong demand for products	− Raw material supply non-existent Processing capacity	− Weak demand	+ Primary products available as inputs
							Relocation to exploit primary products Initial processing possible locally (canning)	

electronics. By contrast, the Japanese firms combine the two benefits and find local markets for part of their production in the newly-industrialised countries, giving added impetus to the competitiveness of the firms.

The third case in Table 2.2 concerns endowment in skilled labour. Here again a distortion may be created between the firm's requirement for skilled labour in order to remain competitive and the country's capacity to supply such labour. The firm relocates to a country where skilled labour is plentiful and/or where the possibilities for the workforce to acquire skills are greater.[12] This relocation can involve just human capital, which will then receive or complete its training in the country with the higher endowment of skilled labour.[13]

The market for skilled labour is often an internal matter for the firm, which itself looks after the advancement of its own people. Differences between factor supply and demand can emerge directly between two firms in the same sector but in different countries, one being well endowed and the other experiencing a shortage. There is potential for co-operation between these two firms by, for example, the transfer of technology or engineers, taking all the participatory forms already indicated. The firm in the advanced country then supplies its specific factor to the other firm. Their respective endowments in the factor (skilled labour) are the reflection of the general capacities of their respective countries.

With respect to capital endowment, the situations are identical to the previous ones. Either the firm has a potential demand for financial capital or for capital goods which it cannot satisfy in its own country, or, on the contrary, it has capital which is specific to its own sector. It can then meet the demand for these factors from foreign firms unable to find them on their own markets or incapable of acquiring them on their own. In the first case, the investments go towards a 'higher' country for the sector in question; in the second case, towards the 'lower' country. The direction of the relationship will be determined by the strength or weakness of the competitive forces of the respective firms. In a given sector, some will go abroad to acquire what they lack; others will wait to get foreign capital to meet their needs. An identical situation will thus create increasing intra-sectoral investment, with the relative dynamism of the firms the deciding factor. Other firms which go abroad are those which have become leaders in their sectors after receiving capital from abroad. The two types of flow can nevertheless exist side by side for the same firm. Samsung, again, invested in the USA while at the same time concluding a joint-venture agreement with Hewlett-Packard for building a computer plant in Korea (*Economist,* 1984).

The two final cases concern raw materials, either minerals or agricultural products. The situations of the host countries often take the form of absolute advantages, associated with the substantial availability of supply of these factors or of the intermediate products. There is little domestic demand, on the other hand, whereas it is strong in the investor's country of origin. Here again there is a discrepancy between the capacities of the firms and those of the countries. The MNEs in the investor's country have the technical and other capacities needed to exploit these resources, for which the supply is located abroad. Relocation is unavoidable. The competitive advantage of the MNE may have been generated by gradual learning over time and by the acquisition of know-how in the country of origin through the exploitation of natural resources which have been exhausted or seriously depleted.

For example, Eastman (1982) points out that the large Canadian MNEs operating in raw materials (paper, timber, minerals) were influenced by Canada's relative initial abundance of these resources. The technical knowledge acquired by these firms was then exploited abroad, when the comparative advantages of the countries for the goods developed in favour of the host country and against the country of origin.[14] Identical arguments can help in understanding the behaviour of MNEs *vis-à-vis* vegetable raw materials. In both cases, part of the transformation of the raw material into the semi-processed product can also take place locally.

The other determining factors, especially those associated with economies of scale, can also be covered by the same explanatory diagram.

CONCLUSION

The integrated analysis puts alongside each other the determining factors in all international exchanges — goods, technologies, and direct investments. In reality these determining factors are more or less important, depending on the nature of the product and the economic characteristics of the partner countries; for merchandise trade, for example, econometric tests indicate a greater explanatory power for one determining factor or another, according to its characteristics (Mucchielli and Sollogoub, 1980, ch. 2).

The choice among different forms of trade with other countries and among various possible manufacturing sites will be made on the basis of the size of the discrepancies between the competitive advantages of the firm, which must remain predominant if the firm wants to maintain its competitiveness for a product, and the comparative advantages of the country of origin and the host country. The firm transfers production abroad, while at the same time continuing production or branching out in a new direction on its home territory; the country, for its part, can extend its comparative advantages by means of its growth path. All types of advantage must therefore be seen in dynamic terms. The integrated analysis then leads to the concept of relative stratification of the kind which has been illustrated here.

NOTES

1. Translated and reprinted by permission from Jean-Louis Mucchielli, (1985), *Les Firmes Multinationales: mutations et nouvelles perspectives,* Paris: Economica.

2. Armington (1969) had already suggested that consumers have different appreciations of identical goods produced in different countries.

3. Katrak (1974) found in empirical tests that economies of scale were a powerful explanation of US export performance.

4. These analyses lead to explanations of intra-sectoral trade in the developed countries.

5. For a full examination of this question see Magee (1976). See also the pioneering work of Ohlin (1931).

6. Magee (1976) prefers the term 'differential' to the term 'distortion', regarding the latter as too normative.

7. This reintroduces the analogy between the neo-technological and neo-factorial approaches to trade theory, but the status of the technology is different here: in the last case, it appears simply as a third factor.

8. See Sharpston (1975). The FRG is seen to have a greater abundance of human capital (supervision) than its European partners and of unskilled labour (immigrants) than the USA.

9. Know-how and cost factors also come into this form of relocation.

10. According to Couffin (1977, pp. 45 and 48) 'the attraction of the American market emerges very clearly as the fundamental motive (for French firms to set up in the United States), both because of its size and its growth prospects'.

11. The notion of specific human capital is to be found, for example, in Becker (1964).

12. In the case of the Samsung direct investment in California which was noted earlier, the closeness to the US labour market allowed the firm to recruit 50 local engineers to carry out its R&D (*Economist*, 1984).

13. Grubel and Scott (1966). This can give rise to the 'brain drain' afflicting certain developing countries.

14. The paper by Rugman in this volume (Chapter 8) is also of interest in this regard.

3

Trends in Technological Competitiveness within the OECD, 1970–80[1]

Bernadette Madeuf

INTRODUCTION

The trends in the technological positions of OECD countries during the 1970s can be analysed at two different levels: that of the production of technology and that of its international dissemination. At both these levels, analysis is based on input and output indicators of science and technology (S&T) which are now generally accepted. It should be noted from the outset that there is no unique measure in this field which is meaningful for all the different economies. This is why several variables are looked at in parallel.

Most of this paper is devoted to presenting the changes in technological competitiveness in the strict sense of the term, that is to say, limited to the international dissemination of technology. Concerning the production of technology, there will be no more than a brief reminder of the main tendencies in order to give an indication of the relative situations of the different countries.

PRODUCTION OF TECHNOLOGY

Two kinds of indicator can be used to measure the production of technology: indicators of input (R&D expenditure, numbers of research workers) and of output (patent statistics).

Overall, the research effort has picked up again after slowing down between 1971 and 1975 and has even been accelerating from 1979 onwards. This recovery has been accompanied by an increased concentration in the seven largest countries (USA, Japan, Germany, France, UK, Italy, Canada), which accounted for 92 per cent of R&D expenditure in 1982. There has been a tendency for the share of R&D carried out by private firms to increase, as a result of a more marked acceleration in this type of expenditure after 1975.

Within this overall growth, Japan stands out as being particularly dynamic. Average annual R&D growth was twice the OECD average in the period 1971–81. Japan has strengthened its hold on second place with 17 per cent of Business Enterprise R&D expenditure (BERD) in the OECD region in 1981, compared with 11.9 per cent in 1971. The other notable change, in relative terms, is the decline in the US share from a little over 50 per cent in 1971 to 46.3 per cent of total OECD R&D in 1981. There were some other minor modifications in relative positions (slight progress in the cases of Germany, France, Italy and Sweden; slight declines for the UK, Canada, the Netherlands and Switzerland).

In parallel with the change in the distribution of the R&D effort, there have been substantial changes in the patent figures. There is a sharp contrast between the advance made by Japan in terms of national applications for patents and the slowdown from 1968–70 onwards in all the other countries except the USA. Japan has become the country with the largest number of applications in the OECD area (40 per cent of the 1982 total), followed by the USA with 19 per cent. This rise is due to the increase in the number of domestic applications, with foreign applications playing only a minor role. Unlike the situation in most other countries, the share of domestic applications in Japan is very high (87 per cent) and this is no doubt linked to the special nature of the Japanese system, which is one of 'single claim' patents.[2] It seems useful therefore to supplement this initial conclusion by other patent statistics. The breakdown of patents granted and of foreign applications shows that Japan's performance is somewhat less brilliant. On the basis of these two indicators, the USA retains its first place throughout the period, followed by Germany.

The introduction of international systems such as the EPC[3] and the PCT[4] has altered the conditions under which residents of signatory countries or third countries file patents abroad. The overall result has been to increase the patent applications filed from and in foreign countries, signifying greater internationalisation of patents. Within this general movement, Japan and the USA are in a special situation by comparison with the European countries, especially the smaller ones. The smaller the country the greater the impact of opening it up to foreign applications. The USA and Japan are characterised by a certain degree of resistance to foreign applications, accompanied by an increase in their own applications abroad aided by the new systems. The major European countries like Germany, France and the UK are experiencing an increase in both foreign and external applications.

Possible routes for the international dissemination of technology

In order to examine the situations of the different countries in terms of the international dissemination of technology, two indicators are available classified as 'output' indicators for the S&T system. These are technological receipts and payments, and exports and imports of goods with a high technological content.

Before considering these indicators it should be noted that they correspond to two different forms of international dissemination of technology: the sale of technological information and the sale of goods incorporating the technology. However, a firm with technological assets has a third option: instead of selling licences or exporting, it can set up a network of subsidiaries or make a direct investment. Moreover, once the private firm has a network of subsidiaries, the conditions governing the choice between exporting, granting a licence or transmitting innovations through the subsidiaries will change. Recent studies show that firms generally opt for dissemination through their subsidiaries. It must be understood that this preference has an impact on international flows, as these are usually measured at national frontiers; relocated production takes the place of exports and may give rise to the payment of royalties. There is also the fact that the tendency for firms to become multinational is all the more marked in industries where R&D and technological competition are more decisive.

TRENDS IN TECHNOLOGICAL PAYMENTS

Principal features

Without repeating in detail here the comments to be made concerning the technological balance of payments (TBP) as an instrument for measuring the international dissemination of technology, they can be summarised under two headings: the recording of payments and receipts in the TBP and the economic significance of the data themselves (Madeuf, 1984).

The two main problems as regards the recording of flows are the heterogeneous contents and non-comparability at the international level. TBPs have a heterogeneous content in that they record side by side not only flows relating to the transfer of technology proper (patents, manufacturing licences, know-how) but in some countries also services of a technical nature (assistance, training, consultancy work) and sometimes even factors related to industrial and intellectual property but with no direct relationship to technology (trademark licences, film rights, management services, etc.). Non-comparability at the international level stems not only from the differences in national TBP coverage noted above, but also from variations in the survey procedures and in the way the information is presented.

Problems of interpretation are obviously raised not only by the mixed contents of the TBPs but also by the elusive character of certain international flows of technological knowledge, i.e. those for which there is no visible form of payment. Some examples are cross-licensing, transfer of knowledge to a subsidiary, and international co-operation of a non-commercial type. Another factor is the behaviour of the firms mainly responsible for transferring technology, i.e., multinational enterprises, for whom the type of payments registered in the TBP may be only one of several possible channels of reimbursement for technology transferred to subsidiaries. Their choice between these channels will be affected by fiscal and other considerations which may lead to the TBP data seriously overestimating or underestimating the real flows of technology involved. This means it is difficult to make an economic interpretation of intra-firm flows on the basis of accounts which, in the last analysis, are based on the firms' worldwide strategy.

Overall trends

At present the OECD has collected TBP information for only 15 countries.[5] A complete set of figures for the period 1972–82 is available for only eight of them. However, these represent 80 to 90 per cent of the total, as they include the six most important countries for payments and receipts (USA, Japan, Germany, France, UK, Italy in 1972, the Netherlands in 1982).

The first general tendency to be noted is the concentration of receipts and payments in a limited number of countries. The second concerns the trend in receipts and payments of the eight countries

Figure 3.1: Technological balance of payments for eight OECD countries

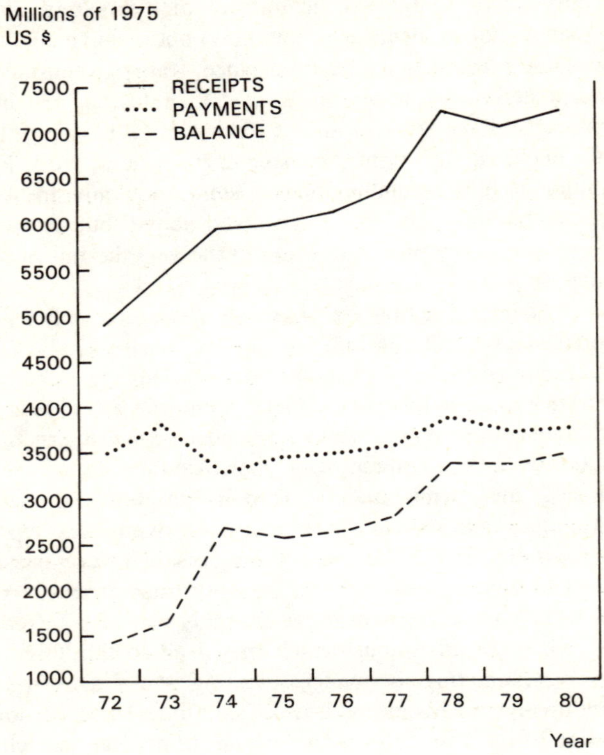

Source: For Figure 3.1 and tables, see Note 1 to this chapter.

shown in Figure 3.1. The relative stability of payments contrasts with the growth of receipts, which showed an upward tendency interspersed with periods of acceleration (before 1974, between 1976 and 1978) and slowdown (between 1974 and 1977).

The contrasting trends in receipts and payments lead to a suggestion concerning the geographic distribution of the receipts. Payments by OECD countries are received by other OECD countries, but part of the receipts are by newly industralising countries. The divergency between payments and receipts can therefore be ascribed to payments for transactions involving the dissemination of technology in the direction of the developing countries, with a resulting increase in their share of total receipts.

Country positions with regard to the international dissemination of technology

Payment/receipt shares and balances

The ranking of the eight countries under consideration and the relevant trends are shown in Table 3.1 in terms of receipts and payments, balances and ratios of receipts to payments.

The first point to note is the exceptional situation of the US economy as regards the balance (and ratio), as well as the share in receipts. In spite of a slight decline in the share of receipts, the US is still undoubtedly the principal centre for the dissemination of technology. Among other major OECD countries, only the UK shows a positive balance, but neither this nor its ratio is commensurate with that of the US.

In terms of relative shares of receipts and expenditures, the most significant changes are: the increase in the receipts of Japan and France; the fall in UK receipts; the increase in the payments of France and Germany; and the fall in the payments of Japan, the UK and the US. In sum, Japan and Germany have switched rankings as centres for the dissemination of technology (4th and 5th).

If the positions of the various countries as regards balances or ratios of receipts to payments are examined, there are various cases to be noted:

(a) in Japan, the USA and the UK, ratios have improved due to increased receipts *and* lower payments, the trend in Japan being towards reduced dependence and a larger role in dissemination;
(b) an improved ratio is associated with either a stagnation in flows (UK) or increased flows (France);
(c) slight improvement in the ratio due to stronger increases in receipts than in payments (Germany), or the reverse (Austria);
(d) the Netherlands has a worsening ratio owing to a more marked increase in payments.

Technological dependence and competitiveness

The scale of the various countries' technological payments and, more particularly, their receipts is related to their size and

Table 3.1: Technological balance of payments (millions 1975 US dollars)

	1972						1982					
	Receipts R	%	Payments P	%	Balance B	Receipts/ Payments %	Receipts R	%	Payments P	%	Balance B	Receipts/ Payments %
USA	3,214.8	(66.7)	368.3	(10.0)	2,846.4	873.0	4,130.9	(63.3)	202.4	(5.4)	3,928.5	20.4
Japan	226.5	(4.7)	934.0	(25.5)	−707.5	24.2	526.5	(8.1)	804.7	(21.6)	−278.2	65.4
FRG	270.4	(5.6)	631.6	(17.2)	−361.1	42.8	342.1	(5.2)	679.5	(18.3)	−337.4	50.3
France	302.8	(6.3)	460.8	(12.6)	−158.0	65.7	552.4	(8.5)	644.2	(17.3)	− 91.9	85.7
UK	563.9	(11.7)	511.0	(13.9)	52.9	110.3	610.9	(9.4)	463.6	(12.5)	147.2	131.8
Italy	81.3	(1.7)	472.8	(12.9)	351.5	17.2	133.9	(2.0)	498.6	(13.4)	−364.8	26.9
Netherlands	151.3	(3.1)	223.1	(6.1)	− 71.8	67.8	210.0	(3.2)	352.7	(9.5)	−142.8	59.5
Austria	12.0	(0.02)	63.8	(1.7)	− 51.8	18.8	20.9	(0.3)	73.0	(2.0)	− 52.1	28.6
Total	4,823.0	(100)	3,665.4	(100)	1,157.6	131.6	6,527.6	(100)	3,718.7	(100)	2,808.5	175.5

technological potential. Without attempting to establish correlations calling for very lengthy analysis, a relationship can be said to exist between receipts and payments and R&D expenditure, although it is not a symmetrical one.

The purchase of foreign technology, and the related payments, can be treated as equivalent to an import of technology supplementing national supply. The ratio of technological payments to BERD (to be denoted by M) may serve as a yardstick of dependence on external sources of technology. If, similarly, technological receipts are equated to an export of technology, the ratio of receipts to BERD (to be denoted by X) provides a rough estimate of a country's technological competitiveness. Calculations of these ratios also enables the absolute and relative data to be weighted by the size of R&D budgets and so reduce the effects of the differences in size of national economies and technical potential.

Tables 3.2 and 3.3 show the values of these two ratios for the years 1969 and 1979 in 15 OECD countries. The first conclusion to be drawn is that the five major countries responsible for the bulk of the flows recorded in the TBP have fairly low figures for X and M (less than 20 per cent). In other words, international movements play a relatively minor role in either purchases or sales of technology. The same is true of Sweden. Within this first group, Japan and Germany have reduced their dependence on foreign technology; on the other hand, they seem to export less than the three others (France, the USA, the UK) whose export ratios show a tendency to increase.

A second group of countries consists of those where the ratios fall between 25 and 40 per cent of BERD. With the exception of Denmark and Finland, they all import more than they export. A third group is made up of two countries, Portugal and Spain, whose dependency ratios on foreign technology are over 100 per cent, owing to their levels of industrialisation and the weakness of their technological potentials.

Form and content

In addition to the aggregated data analysed above, it would be useful to have additional qualitative information on the forms and content of technological dependence or competitiveness, especially concerning two particular aspects.

The first is the composition of technological receipts and payments. It was pointed out earlier that the balances are not homogeneous and that technical assistance, studies, etc. are often

Table 3.2: Index of technological dependency, 1969 and 1979 (BERD = 1)

Countries by declining rank following the 1979 index	1969	1979	
1. Portugal	1.37[e]	2.36[d]	
2. Spain	2.37[f]	2.17[b]	
3. Finland	0.36	0.55	
4. Australia	—	0.38[d]	
5. Netherlands	0.30	0.36[b]	
6. Italy	0.41[e]	0.33	
7. Denmark	0.34[a]	0.33[c]	
8. Austria	0.49	0.32[d]	
9. Canada	0.26	0.28	
10. France	0.15[a]	0.16	
11. UK	0.14	0.14[d]	
12. FRG	0.16	0.11	(0.14)
13. Japan	0.17[e]	0.08	
14. Sweden	0.06	0.03[c]	
15. USA	0.01	0.02	

Notes: a. 1970 b. 1976 c. 997 d. 1978 e. 1972 f. 1973
BERD stands for Business Enterprise Research and Development.
— Data not available

Table 3.3: Index of technological competitiveness, 1969 and 1979

Countries by declining rank following the 1979 index	1969	1979	
1. Finland	0.24	1.27	
2. Denmark	0.42[a]	0.41[e]	
3. Spain	0.26[c]	0.28[d]	
4. Portugal	0.11[b]	0.25[f]	
5. Netherlands	0.24	0.21[d]	
6. UK	0.13	0.17[f]	
7. USA	0.11	0.17	
8. France	0.09[a]	0.14	
9. Italy	0.07[b]	0.11	
10. Canada	0.06	0.08	
11. Austria	0.10	0.07[f]	
12. Australia	—	0.06[f]	
13. Japan	0.03[b]	0.05	
14. FRG	0.06	0.04	(0.05)[g]
15. Sweden	0.06	0.03[e]	

Notes: a. 1970 b. 1972 c. 1973 d. 1976 e. 1977 f. 1978
g. 0.14 in 1977. The German data related to BERD in 1979 cannot exactly be compared to those of early years as the coverage has been larger in more recent surveys.

included alongside the patents and licences corresponding to technological transfers in the strict sense. An analysis of the balances of the various countries, distinguishing between patents and licences on the one hand and technical assistance and studies on the other would provide useful information on the subject of dependence and competitiveness. France is a case in point (OECD, 1984). From 1970 to 1982 the TBP showed a steady overall negative balance in current prices and a decreasingly negative one in constant prices. However, this relative improvement masked a substantial deficit as regards patents and licences, offset by increased receipts for technical studies. The German TBP has similar characteristics.

The second structural factor concerns the participation of multinational enterprises in international flows of technology and the relevant payments. Few countries provide a breakdown by type of ownership of firms. The OECD has as yet only been able to compile data for the USA and the UK and to some extent for Germany.[6] Analysis of the trends over 15 years (1967–82) shows that in the USA and the UK the share of receipts earned by the subsidiaries of multinational companies grew significantly. In the other direction, payments by subsidiaries of foreign firms were recorded in all three countries (see Table 3.4). A wider international comparison would no doubt show interesting changes in this respect.

EXCHANGES BETWEEN HIGH R&D-INTENSITY INDUSTRIES

Main features

One way in which technology is diffused internationally is by being incorporated in goods. It would seem plausible that products with the highest new technology content would come from industries with the most substantial R&D efforts. Furthermore, the degree to which a country is successful in exporting such products could be used as an indicator of its international technological position.

Although these ideas seem attractive at first sight, difficulties arise as soon as we start to try and define exactly which activities are to be analysed. There are no clearly agreed concepts but rather a multiplicity of rather similar terms, such as ‘high technology’, ‘advanced technology’, ‘core technology’, or ‘strategic technology’, all of which are based on the idea that R&D and technological mastery contribute to industrial and commercial success.

Table 3.4: Share of income and payments corresponding to flows between affiliated companies (in %)

	Income					Payments				
	1967	1970	1975	1980	1982	1967	1970	1975	1980	1982
USA	74.1	73.1	81.1	82.1	–	37.3	49.3	60.1	67.5	–
UK	35.3	33.2	31.8	43.2	52.3	58.1	59.1	65.1	84.8	87.6
FRG[a]	2.4	3.6	5.1	8.1	19.9[b]	65.3	59.4	67.4	61.7	67.9[b]

Notes: a. Differing from the data for the US and UK companies, incomes include those of German companies with partial foreign ownership. Only the payments can be strictly compared.
b. 1983.

The characteristics which are usually attributed to activities considered as being in the high-technology class are the following: the need for a strong R&D effort; the strategic importance for governments; very rapid product and process obsolescence; high-risk capital investment; and a high degree of international competition in R&D, production and marketing.

One criterion eventually emerges as being really specific to high technology, namely, the intensity of the R&D effort, with the other criteria being applicable to other activities. As a result, the definition of technological intensity can be reduced to that of R&D. Industries with high R&D effort will therefore be considered as industries with high technological intensity.

This paper uses the approach by industry rather than that by product which was adopted in an earlier study (OECD, 1983). The level of aggregation is that used in the OECD's international surveys on personnel and expenditure devoted to R&D (OECD, 1981). Technological or R&D intensity is measured by the ratio of R&D expenditure to production. This ratio is calculated for each industry for the eleven reference countries taken together as an area.[7] It is an average weighted by each industry's share in total output for the eleven countries.

The industry rankings for the years 1970 and 1980 are given in Table 3.5. Six industries are in the high-intensity category with a growing average coefficient due to the computer and electronics industries. By definition these six industries are responsible for most of the R&D expenditure in the area, accounting on average for 51 per cent of the total in the period 1970–80. Their average share over the same period, however, is much lower in terms of production (11 per cent) or of exports (16 per cent), as noted in Table 3.6. The concentration of R&D in the high R&D-intensity industries turns out to be especially important in four countries: the USA (63.6 per cent), the UK (62.8 per cent), France (57.5 per cent) and Germany (51.6 per cent). It is lower in Japan where it was 40 per cent in 1981.

Although the high R&D-intensity industries account for only 16 per cent of industrial exports, they have two major features which should be noted:

(a) comparison between the three groups of industry shows that the relative propensity to export is higher on average for the high R&D-intensity group (1.45, as against 1.37 and 0.77 for the medium- and low-intensity groups, respectively);
(b) there is a positive correlation between the share in manufactur-

Table 3.5: Intensity of R&D expenditure in the OECD area (weighting of the 11 main countries — R&D expenditure/output)

1970	Intensities	1980	Intensities
High		High	
1. Aerospace	25.6	1. Aerospace	22.7
2. Office machines, computers	13.4	2. Office machines, computers	17.5
3. Electronics and components	8.4	3. Electronics and components	10.4
4. Drugs	6.4	4. Drugs	8.7
5. Instruments	4.5	5. Instruments	4.8
6. Electrical machinery	4.5	6. Electrical machinery	4.4
Average	10.4	Average	11.4
Medium		Medium	
7. Chemicals	3.0	7. Automobiles	2.7
8. Automobiles	2.5	8. Chemicals	2.3
9. Other manuf. ind.	1.6	9. Other manuf. ind.	1.8
10. Petroleum refineries	1.2	10. Non-electrical machinery	1.6
11. Non-electrical machinery	1.1	11. Rubber, plastics	1.2
12. Rubber, plastics	1.1	12. Non-ferrous metals	1.0
Average	1.7	Average	1.7
Low		Low	
13. Non-ferrous metals	0.8	13. Stone, clay, glass	0.9
14. Stone, clay, glass	0.7	14. Food, beverages, tobacco	0.8
15. Shipbuilding	0.7	15. Shipbuilding	0.6
16. Ferrous metals	0.5	16. Petrol refineries	0.6
17. Fabricated metal products	0.3	17. Ferrous metals	0.6
18. Wood, cork, furniture	0.2	18. Fabricated metal products	0.4
19. Food, beverages, tobacco	0.2	19. Paper, printing	0.3
20. Textiles, footwear, leather	0.2	20. Wood, cork, furniture	0.3
21. Paper, printing	0.1	21. Textiles, footwear, leather	0.2
Average	0.4	Average	0.5

ing exports taken by high R&D-intensity industries and the R&D intensity of those industries. This correlation was statistically significant when tested for the eleven countries shown for 1970 and 1980.

It therefore seems justifiable to use the data concerning the exports of this group of industries to analyse the positions of the various countries in the international dissemination of technology.

Table 3.6: Average weights of the manufacturing industries of the OECD area (11 countries) during the period 1970–80

	R-D	Output	Exports
High intensity	51%	11%	16%
Medium intensity	32%	32%	44%
Low intensity	17%	57%	40%

COUNTRY POSITIONS

A country's position in international trade by high R&D-intensity industries can be studied from various viewpoints. Three which are of interest are the importance of high R&D-intensity industries in manufacturing output and exports, the degree of specialisation, and apparent comparative advantage.

Manufacturing output and exports (Table 3.7)

A comparison of the shares of high-intensity industries in manufacturing output and exports in each country gives a number of pointers. The USA specialises the most in this type of exports. Its share has even increased, although the opposite occurred for output. The USA appears to have a relative advantage in high R&D-intensity industries, probably due to its large R&D efforts (63.6 per cent of the total, the highest proportion in the OECD area).

A second finding is that the shares of high-intensity industries in exports increased faster than their share in output. This denotes greater emphasis on trade, consistent with the previous remarks on the propensity to export. With the exception of the Netherlands, the export share of output has risen; this is particularly the case in the USA, the UK, Japan and Canada.[8] The third point to be noted is that in 1980 the weight of high-intensity industries in exports in the five major countries (USA, Japan, UK, France and Germany) was higher than the OECD average, which was 16 per cent for the 1970–80 period. We shall concentrate mainly on these countries.

Table 3.7: Weights of high R&D-intensity industries in total manufacturing output and exports

	Output			Exports		
	1970	1975	1982	1970	1975	1982
USA	14.6	12.5	10.8	25.8	24.6	31.1
Japan	14.1	12.2	13.4	20.2	17.2	26.9
FRG	11.9	12.1	12.0	15.5	14.6	17.7
France	10.3	11.1	11.3	13.9	13.6	18.2
UK	12.2	11.2	12.5	16.8	18.8	24.8
Italy	11.7	11.2	12.1	11.5	9.8	12.3
Canada	8.5	7.5	6.7	8.8	7.7	10.0
Australia	7.2	8.3	7.6	2.7	4.4	3.7
Netherlands	11.8	12.3	12.3	15.9	14.1	12.8
Sweden	9.6	10.3	10.0	11.7	12.9	14.6
Belgium	6.5	6.8	6.6	7.1	8.5	8.8
EEC	11.4	11.3	11.7	15.9	15.5	16.8

Balances, specialisation by 'niche' and intra-industry specialisation

The trade balances of the high R&D-intensity industries as a whole are positive in the five major countries only. This result should be interpreted in the light of the size of the domestic market in these economies, for the biggest countries can specialise in several of these industries while the smaller countries must restrict themselves to only one or two. In structural terms, it is therefore more difficult for the latter countries to achieve a positive trade balance in their high R&D-intensity industries as a whole.

The information provided by the balances is supplemented by two specialisation indicators. They concern:

(a) Specialisation by 'niche' which is defined as the ratio of output to domestic demand. The higher the value of this indicator, which measures the surplus of output over domestic demand, the greater is the country's specialisation (positive balance). Values of around 100 mean that specialisation is balanced.

(b) Intra-industry specialisation which is given by the ratio $(X-M)/(X+M)$. This ratio ranges from $+1$ (exclusively exporting country) to -1 (exclusively importing country). It takes into account the total volume of trade. A big surplus $(X-M)$ is required for the ratio to be well above 0 for a country heavily involved in trading (substantial $X+M$).

Table 3.8: Specialisation by 'niches', industry specialisation and apparent comparative advantage in higher R&D-intensity industries

	S.B.N.[a]			I.S.[b]			A.C.A.[c]		
	1970	1975	1980	1970	1975	1982	1970	1975	1982
USA	105	110	109	0.37	0.40	0.18	156	156	152
Japan	109	114	126	0.44	0.53	0.64	122	108	131
FRG	118	121	115	0.29	0.27	0.14	94	92	86
France	100	104	102	0.00	0.09	0.04	84	86	89
UK	108	113	108	0.18	0.18	0.02	101	119	121
Italy	99	99	94	−0.00	−0.01	−0.01	70	62	60
Canada	81	73	66	−0.27	−0.41	−0.03	53	49	49
Australia	61	66	64	−0.83	−0.74	−0.82	16	28	18
Netherlands	93	100	97	−0.04	0.00	−0.03	96	89	62
Sweden	89	97	95	−0.13	−0.02	−0.04	71	81	71
Belgium	77	81	79	−0.17	−0.12	−0.09	43	54	43
EEC	106	110	104	0.10	0.12	0.05[d]	85	87	83[d]

Notes:

a. $\text{S.B.N.} = 100 \times \frac{\text{Production}}{\text{Domestic demand}} = 100 \times \frac{\text{Production}}{\text{Production} - X + M}$

b. $\text{I.S.} = \frac{X - M}{X + M}$ X : Exports
M : Imports
C : X/M

$= \frac{C - 1}{C + 1}$

c. $\text{A.C.A.}_{ij} = 100 \times \frac{X_{ij}}{\Sigma_i X_{ij}} \Big/ \frac{X_{im}}{\Sigma_i X_{im}}$

d. 1980.

Table 3.8 shows the trend in the values of these two indicators. Only the five countries with a positive balance and the EEC had a 'niche' specialisation indicator consistently above 100. In 1980 specialisation seemed to be greater in Japan and Germany than in the USA. It was on the increase in many countries and in the EEC between 1970 and 1975. This uptrend persisted only in Japan after 1975, in contrast with all other countries.

The intra-industry specialisation ratio for the period as a whole is positive for the same countries and the EEC. The most important trends concern Japan, showing the largest relative surplus and steepest uptrend over time, and the decline in intra-industry specialisation in the other major countries except France, where the ratio remained quite low.

If these performances are converted into export/inport ratios, the following were the front runners in the trade of high R&D-intensity industries in 1982:

	Export/Import Ratio
Japan	4.7 or 470%
USA	1.4 or 140%
EEC	1.1 or 110%

Apparent (or revealed) comparative advantage

In assessing a country's specialisation, its exports from the industries under review must be compared with total exports in order to take into account the effects of scale. This is why an indicator of apparent comparative advantage — also known as revealed comparative advantage — has been established. Each country's share in high R&D-intensity manufactured exports in the OECD area is weighted by its share in the area's total manufactured exports.[9] A coefficient exceeding 100 indicates a comparative advantage, while if it is less than 100 it means a disadvantage and non-specialisation (see Table 3.8).

Only three of the major countries have a comparative advantage in high R&D-intensity industries: the USA has the highest degree of specialisation despite a decline in the indicator. Japan's marked up-trend after 1975 has given it second place. The UK ranks third, mainly due to the improvement in its advantage between 1970 and 1975. No other country, including Germany (declining advantage) and France, has a comparative advantage in high R&D-intensity industries.

The range of indicators and variables used to analyse the position of countries in international trade involving high R&D-intensity industries lead to four main conclusions. First, despite a decline, the USA still leads the field in the dissemination of technology through the export of goods. Next, Japan ranks second in the world thanks to a sustained uptrend. Third, divergent trends are found in the major European countries. Finally, overall this results in a decline for the EEC as regards high R&D-intensity industries.

CONCLUSION

The tendencies described here raise three questions. Two of these concern the relationship between technological competitiveness and industrial competitiveness. The third concerns the validity of the indicators that have been used.

First, to what extent is Japan's progress in the high R&D-intensity industries due to the effectiveness of the R&D itself and/or to more general aspects of industrial competitiveness? Additional analyses covering the totality of industry, including medium- and low-intensity groups, show clearly that Japan has been progressing in all of them. The USA and the EEC, on the other hand, have seen a decline in their competitiveness, but with varying consequences. Because of its greater degree of market openness and its less favourable specialisation (i.e. less concentrated on growth industries), the EEC position seems to be more under threat than that of the USA.

The second question concerns precisely this point of the generally unfavourable changes in the case of the EEC. For the individual countries, there is a striking increase in market penetration by imports and loss of market shares. These changes do not seem to be explained by any fall in R&D, which has risen quite strongly overall except in the UK. The essential problem seems to be that of the effectiveness of European R&D and its subsequent translation into industrial and commercial performance. The solution is not to be found merely in improved R&D organisation; the drawing up of a common programme to eliminate overlap and to develop effective collaboration is a necessary condition, but not a sufficient one. Strengthening European competitiveness is above all an industrial imperative.

The final question concerns the statistical information being used. This goes beyond any reminder of the complementary nature of many of the indicators. It is much more in the nature of a problem which has only briefly been touched on here: the impact of the existence of multinational firms on the significance of national trade statistics. The internationalisation of production means that national indicators lose much of their value for demonstrating the positions of individual countries. Those positions cannot be measured merely by exports, especially by the R&D-intensive industries dominated by the multinationals. The export figures for a given country should be increased by the volume of geographically-relocated production under the control of firms based in that country. Unfortunately this relocated production is often poorly documented, however essential it may seem to us to include it in any assessment of the competitiveness of each economy.

NOTES

1. This text includes material from the summary report prepared by the author for the OECD and included in 'OECD Science and Technology Indicators', OECD, Paris, 1986.
2. The system has recently been altered.
3. The Munich Convention, EPC, October 1973.
4. International Patent Cooperation Treaty, PCT, June 1970.
5. USA, Japan, Germany, France, UK, Australia, Canada, Italy, the Netherlands, Sweden, Austria, Denmark, Finland, Portugal, Spain.
6. For Germany, only the payments are comparable with the data from other countries.
7. USA, Japan, Germany, France, UK, Italy, Canada, the Netherlands, Sweden and Belgium.
8. Its evolution can be measured by the ratio $\frac{Xi/X}{Pi/P}$

where $\frac{Xi}{X}$, $\frac{Pi}{P}$,

are shares of high-intensity industries in exports and output respectively.

9. As in the formula:

$$Ca_{ij} = 100. \frac{X_{ij}}{\Sigma_i X_{ij}} / \frac{X_{im}}{\Sigma_i X_{im}}$$

X_{ij} : Exports of industry j in country i

X_{im} : Manufacturing exports of country i

4

International Knowledge Transfers and Competitiveness: Canada as a Case Study

D.C. MacCharles

INTRODUCTION

Exports, direct investment and licences are all substitute ways for foreign manufacturers to access Canadian markets. The particular combination of them used by a firm depends mainly on which provides the largest profit and rent in the host country's prevailing economic environment. All three methods provide Canada with knowledge about existing as well as new management and technological practices which stimulate productivity, competitiveness and growth. However, each has different effects and implications for Canada's stock of intellectual capital.

Exports to Canada have knowledge and technical expertise embedded in them which forces domestic producers to operate at a similar level in order to be competitive. In this case, imports by Canadians indirectly contribute to an accretion in the nation's stock of intellectual capital. Direct investment from abroad transfers knowledge and technical expertise to subsidiaries and thereby directly adds to the nation's stock of intellectual capital. Similarly, licences provide a direct transfer of knowledge, but usually to third parties who are often Canadian nationals.

Changes in trade conditions, competitiveness and profitability over the past decade in Canada and abroad have altered the relative use of direct investment, imports and licences. Foreign direct investment has become relatively less important as a means for transferring knowledge, with imports (and licences) becoming relatively more important. These issues of productivity, costs and competitiveness as they interrelate with the transfer of knowledge to Canada's manufacturing sector provide the focus for this paper.

THE HISTORICAL BACKGROUND

Market size

Domestic markets in Canada for manufactured goods are small and fragmented regionally compared to those in the USA, Japan and the EEC. Yet the plants of the larger manufacturers in Canada tend to be near minimum efficient size at about 75 per cent of that for counterparts in the USA (Baldwin and Gorecki, 1983a). Such large plants in relation to the size of the market led manufacturers to increase their rates of capacity utilisation by producing a wide variety of products (horizontal diversification) along with contracting-in a wide variety of semi-finished materials, components and other activities such as services (vertical diversification) compared to plants in the USA (Caves, 1975). The relatively high degree of horizontal and vertical diversification did increase rates of capacity utilisation and reduce unit fixed costs through economies of scope. But it added to unit variable costs by increasing the degree of complexity in the firms and creating product-specific diseconomies of scale (Daly, Keyes and Spence, 1968). These latter cost-increasing influences from diversity more than offset its benefits of increased fixed cost absorption. In addition, over 90 per cent of plants are small by international standards, with fewer than 200 employees (see Table 4.2). These smaller plants have low productivity and high unit costs because of their diseconomies of scale.

Barriers to trade

In order to protect manufacturers from lower-cost imports, a national policy was developed based on the use of various barriers to trade, including high tariffs. These barriers allowed producers to pass on higher costs to customers through higher prices. It also affected corporate decision-making. Managers have been slower in adopting state of the art practices used by producers elsewhere in the world (Daly and Globerman, 1976). More recently, this has shown up in Canadian manufacturers being slower than their foreign counterparts in adopting quality circles, 'just-in-time' inventory control techniques, CAD/CAM methods, robotics, more efficient plant layouts and permanent employment practices (Daly, 1984a; European Management Forum, 1984 and 1985; Palda, 1984).

Protection also helps to explain why the average Canadian manager tends to be older, to have moved through the ranks more slowly, to be less well educated (although this is slowly changing with the increase in the flow of MBA graduates) and to be generally less experienced than managers elsewhere in the world (Daly, 1979). A more competitive environment wuld have required managers to adopt sooner the changes taking place in the rest of the world in technological and managerial practices, with a consequent improvement in productivity and competitiveness.

Unit labour costs

One measure of the competitive disadvantage of Canadian manufacturers is unit labour costs. On an exchange-rate-adjusted basis, Canadian unit labour costs in 1984 were about 15 per cent higher than for the USA and about double those of Japan (Daly and MacCharles, 1986a). This disadvantage reflects to a large extent the build-up of inefficiencies in the manufacturing sector as a result of protectionist government policies and poor management practices, as several studies have shown (Daly, Keyes and Spence, 1968; Daly and MacCharles, 1986b; Harris, 1983; West, 1971; Wonnacott and Wonnacott, 1967).

Increased competition from a freer trade policy would lead to productivity and cost improvements by domestic firms. Imports would increase and force firms to rationalise in order to survive. Rationalisation would be accomplished through increased scale of operations and increased specialisation. In order to increase their scale of production, most Canadian manufacturers must export. This is because the domestic market is small and any increased output could not be absorbed by it without such significant price reductions that scale increases achieved through this alternative would be unprofitable. But to small manufacturers that are price takers in international markets, exports would be into markets with higher price elasticities and profit potential. Increased specialisation requires contracting out the production of finished and semi-finished goods, materials and services to more efficient suppliers both at home and abroad, leading to an increase in imports. Increased imports resulting from freer trade would also force Canadian firms to adopt improved management and production techniques. And imports of components and equipment would indirectly transfer improved technology and knowledge to their Canadian buyers, thereby assisting them in improving their productivity and costs. Imports are particularly helpful to smaller manufacturers in achieving access to state of the art technology since they often lack the depth of organisation needed to stay abreast of all the technical and managerial developments in their industry.

Changing competitive environment

Over the 1970s and into the early 1980s there has been an increase in excess manufacturing capacity worldwide due to slower growth in markets in conjunction with the development of new producers in such countries as Japan and the NICs. Further, since the Second World War, GATT negotiations have resulted in reductions of trade barriers and in generally improved conditions for international trade. Transportation and communication costs have also trended downward. The net result has been increased competition for Canadian manufacturers, particularly since the early to mid-1970s when these factors became strongly cumulative. More recently, the severe recession of 1981–2 significantly reduced demand with a consequent further increase in excess capacity and price competition on a worldwide basis. The result is that since the early 1980s high-cost Canadian manufacturers have faced increased difficulty even in surviving, while unemployment rose to post-Second World War highs.

The 1981–2 recession in Canada was deeper than for almost every other industrial nation and the pattern of change in many of the key economic measures was different than in other business cycles since the Second World War. In particular, the recession was steeper than in the USA (by about 50 per cent measured in GDP terms for the manufacturing sector) and the recovery has been more sluggish. Also, rates of return on investment in the manufacturing sector in both nominal and real terms declined over the 1970s and were close to zero on an inflation-adjusted basis in the early 1980s (Daly, MacCharles and Altwasser, 1982). Real rates of return on investment have subsequently recovered from their recession lows, but are still low in relation to their historical values prior to the late 1970s and those currently prevailing in the USA. In turn, the low rates of return have seriously reduced cash flows available for reinvestment by companies and the profitability of increased capital goods spending. Business investment spending is recovering, but it remains below the level that existed prior to the 1981–2 recession

(Lafleur, 1984; Statistics Canada, 1985a). The increased level of competition for high-cost Canadian manufacturers also shows up in: Canada's declining share of the world market for manufactured goods (Astwood, 1981; Government of Canada, 1983); the large trade deficit on manufactured goods (Britton and Gilmour, 1978); and the declining proportion of world direct investment flows coming into Canada along with an increasing share of outflows (Safarian, 1983; Rugman, 1986).

FOREIGN DIRECT INVESTMENT

Barriers to trade

In addition to creating inefficient resource allocation, Canada's protectionist policies significantly influenced the structure of Canadian industry. One of the more important influences was attracting large inflows of foreign direct investment (FDI). Foreign manufacturers, in order to gain access to Canadian markets, built plants behind the tariff wall since this was cheaper than exporting from a home country and paying the tariff. Also, tariffs were higher on finished than intermediate goods, which increased the degree of effective protection for value added in Canada and made tariff factories worthwhile (both foreign and Canadian-controlled). Foreign-controlled subsidiaries (mostly US) now account for about 50 per cent of total assets in the manufacturing sector, although it varies considerably by industry within the sector.

Knowledge-intensive industries

It is unlikely, however, that trade barriers alone are sufficient to explain the high levels of FDI, since it is not always associated with industries that have high tariffs. The textiles industry is a case in point. While tariffs are important, it is clear that FDI is also associated with industries for which knowledge is a significant input, especially when it is required in indivisible units that are large in relation to domestic market sizes.

Knowledge is an important part of total cost in industries where technologically advanced production methods, advanced management practices and large inputs of R&D are significant factors. Also, the stocks of intellectual capital needed by firms in these industries to produce their knowledge requirements can be a large portion of their total assets. For instance, it has been estimated that for the pharmaceutical industry such stocks are about 30 per cent of total assets (Palda, 1984).

Capturing rent on knowledge capital

The stocks of intellectual capital that some firms have about production processes, products and management practices are usually in the form of intangible assets that become embedded in employees through experience and education (Machlup, 1962). Firms can earn rent on this knowledge capital by exporting. However, when exporting is not feasible for reasons such as barriers to trade, setting-up a subsidiary in a foreign country is an alternative.

These factors explain why the use of FDI has been a favoured vehicle for foreign firms seeking access to Canadian markets. They are in knowledge-intensive industries and the use of direct investment abroad permits knowledge to flow relatively unhindered between the parent and the subsidiary, allowing knowledge to be exploited within the MNEs' sphere of control with a minimum of risk. The intra-firm method of technology and knowledge transfer has been called internalisation because it uses the internal markets of the MNE, rather than external ones, to transfer knowledge and exploit any advantages associated with it (Buckley, 1979; Dunning, 1981; MacCharles, 1978; Rugman, 1981).

An MNE could transfer knowledge directly to third parties by selling it using a licence agreement. But this is usually less desirable than internal transfer to a subsidiary because of the uncertainty and risk associated with allowing third parties access to knowledge which may be proprietary in nature.

It is clear that setting up subsidiaries by foreign firms was a rational response by them to the presence of Canadian tariffs and other trade barriers in industries where intellectual capital is significant and an important cost of production. Subsidiaries have ready access, often at little or no cost, to the knowledge and practices of parents and affiliates (Safarian, 1986). FDI has proven beneficial to Canada as a whole. Productivity benefits arising from the use of the knowledge of MNEs do not stop at their subsidiaries, but tend to 'spill-over' and improve the performance of Canadian-controlled manufacturers as well (Globerman, 1979). Further, the tendency for subsidiaries to be in industries that require substantial inputs of knowledge has meant above-average transfers of it and improved technical progress for the Canadian economy. This is why large inflows of foreign direct investment from the USA have been associated with faster growth in total factor productivity than in periods during which the flows declined. A major part of the productivity growth came about through technological improvement rather than through other influences such as economies of scale (Cacnis, 1985). Also, the ease with which intra-firm transfers of knowledge take place helps to explain why subsidiaries adopt technical changes sooner and why they have higher productivity than their Canadian counterparts (MacCharles, 1981).

Specialisation by subsidiaries

The subsidiaries, especially the smaller ones, tend to manufacture just high-volume items in order to give the major products of their parents access to Canadian markets. They import minor products and components extensively from affiliates in the USA that sell into larger and more competitive markets. The larger size and greater specialisation of affiliates allows them to be low-cost sources of supply to Canadian subsidiaries. This access gives the subsidiaries two advantages over their Canadian-controlled counterparts. First, they can be more specialised both horizontally and vertically and still offer a high level of product variety to customers. Second, the subsidiaries achieve lower costs for products and components. Canadian-controlled firms, especially the smaller ones, tend to use domestic suppliers or else internally produce such items more than do subsidiaries (MacCharles, 1984). This makes them higher-cost manufacturers, compared to the subsidiaries, since production in Canada is generally by small production runs.

KNOWLEDGE TRANSFER AND SMALLER FIRMS

Companies need access to new knowledge about more efficient techniques of production, better management practices and new products if they are to improve their productivity. But they also need access on a continuing basis to existing knowledge in these areas in order to maintain their productivity at prevailing standards for an industry (MacCharles, 1978; Machlup, 1962). While knowledge is a key factor in determining the productivity of a firm, it is not uniformly available to all, especially the smaller Canadian-controlled ones.

Quality knowledge is easily and efficiently transferred to subsidiaries, even the small ones, by parents and affiliates in divisible units at minimum cost of production. Larger Canadian-controlled firms also have the capability to produce or purchase good quality knowledge at minimum cost because of their size and ability to attract personnel from international markets.

But smaller Canadian-controlled firms have a disadvantage in the acquisition and use of knowledge. They may try to produce the complete range required in such areas as marketing, production, accounting and finance thereby incurring the cost penalty of unexhausted economies of scale in its production. Or they may try to get along without some knowledge which also lowers productivity and raises costs. The smaller firms also tend to be restricted to the Canadian market for personnel which has a relatively lower supply of highly trained and educated managers compared to countries such as the USA. The net result is that smaller Canadian-owned firms have access to knowledge that tends to be of lower quality and higher cost thereby lowering overall productivity.

IMPACT OF MNEs ON TRADE FLOWS

The propensity of subsidiaries to contract-out to affiliates does have a significant impact on Canada's imports. The import propensities of subsidiaries in the manufacturing sector tend to be significantly higher than for comparable Canadian-controlled manufacturers (Statistics Canada, 1985b). Most of the difference is attributable to the subsidiaries specialising in the production of just major product lines and importing minor products as wholesalers from affiliates. There is no significant counterpart wholesale activity for Canadian-controlled manufacturers.

Canada's international trade in manufactured goods has been integrated with production data to show the effect of the different import performance between the sectors of control (Table 4.1). Canadian imports, exports and shipments for manufactured goods were all coded to the plants responsible for these activities, using the same product classification concordance for all three of them. The plants within an industry were then assigned to one of the following four trade groups, which classified them by their type of involve-

Table 4.1: Imports, exports and shipments of manufactured goods by manufacturers by sector of control and trade category of enterprises, 1979

Trade category of enterprises	Foreign sector $000,000		Foreign sector % of total		Canadian sector $000,000		Canadian sector % of total	
	Mpts	Xpts	Mpts	Xpts	Mpts	Xpts	Mpts	Xpts
Importers only	$8,517	—	23%	—	$3,691	—	10%	—
Exporters only	—	$2,112	—	6%	—	$3,495	—	9%
Import and export	20,459[a]	19,307[b]	59	54	2,758	11,277	8	31
Total	28,976[a]	21,419[b]	82	60	6,449	14,722	18	40
Shipments	66,627[c]	60,810[c]	59	55	46,073	49,362	41	45

Notes:
a. includes autos and auto parts worth $13,333.
b. includes autos and auto parts worth $9,073.
c. includes autos and auto parts worth $14,422.

Source: Adapted from MacCharles (1984).

ment with international trade: importer only; exporter only; both importer and exporter; and neither an importer nor an exporter. The plant level data were then aggregated to 159 industries by sector of control.

The relatively large foreign sector of control, the high unit costs of Canadian suppliers, and the greater specialisation and wholesale functions of subsidiaries, resulted in them dominating the value of imports by Canadian manufacturers. About 80 per cent of manufactured imports are by the subsidiaries, even though they represent only about 55 per cent of shipments. Further, it is estimated that about 80 per cent of imports by the subsidiaries are intra-firm, with about two-thirds of them due to wholesale activities (Helleiner, 1979; Helleiner and Lavergne, 1979; MacCharles, 1984).

The subsidiaries utilise their foreign affiliates also as wholesalers to access export markets. This, in conjunction with their imports from affiliates, results in a large two-way flow of goods in similar products. Such two-way trade in similar products by firms in an industry is called intra-industry trade (IIT) to distinguish it from the one-way trade normally associated with the traditional models of trade. IIT has facilitiated the rationalisation of the subsidiaries into the North American production and marketing systems of their parents, a process that increased as competitive pressures increased through the past decade or so. Also, given the important role of subsidiaries in manufacturing, their two-way trade explains why Canada has one of the highest elasticities of exports with respect to imports in manufactured goods of any of the industrialised nations (Aquino, 1978). As rationalisation through specialisation takes place, the two-way flow of similar goods increases. There is an increase in imports of lower cost items from affiliates and an increase in exports.

IIT is not as widespread as might be desired, given that it indicates an efficient response to increased competition and an improvement in the technology and knowledge used by subsidiaries in Canada. However, it is more extensive in the foreign than Canadian sector of control. As an indication of the extent of horizontal specialisation, in the foreign sector of control in 1979 only 32 out of 159 industries had a significant level of intra-industry trade in finished goods of major product lines (MacCharles, 1986). The figure was even lower for the Canadian sector, at 18 industries. The extent of vertical specialisation is evident from the fact that the number of industries actively involved in IIT increased to 54 for the foreign sector of control and to 39 for the Canadian sector when trade in intermediate goods was taken into account.

Table 4.2: Selected comparisons between sectors of control by plant size, manufacturing sector, 1974

Plant size measured in employees	Value-added (1) production worker (ratio Cdn. to Fgn.)	Percentage of plants: Cdn.	Fgn.	Percentage of sales: Cdn.	Fgn.
Fewer than 50	0.50				
50 to 200	0.67	88%	6%	19%	5%
200 to 400	0.75				
Greater 400	1.00	2%	4%	23%	53%

Source: Adapted from D.C. MacCharles, 'The Performance of Direct Investment', and Statistics Canada Publications.

IMPACT OF MNEs ON CANADIAN PRODUCTIVITY TO THE MID-1970s

Table 4.2 shows the value added per production worker for different plant sizes in 1974, with the value for the Canadian sector of control expressed as a ratio of that for the subsidiaries. Plant size is expressed in terms of the number of employees working in a plant. The proportions that the plants in each size category, and the sales of these plants, represent of total plants and sales in the manufacturing sector are also shown.

The larger Canadian plants tend to have levels of value added per employee broadly comparable to foreign-controlled organisations in the same industry and size group. This result is broadly similar to another study (Safarian, 1966). The smaller subsidiaries, however, tend to have significantly higher levels of value added per employee than the Canadian-controlled plants in the same industry and size groupings. Further, the smaller subsidiary plants and firms tend to have levels of value added per employee rather similar to the larger units, indicating there are only moderate economies of scale at the plant level for the subsidiaries. This suggests there is a significant impact on the productivity of subsidiaries (especially smaller ones) because of intra-firm transfers of knowledge, greater horizontal and vertical specialisation and intra-firm trade in goods (MacCharles, 1981).

Reduced impact of MNEs on Canadian productivity after the mid-1970s

The increased competition from foreign producers over the past decade or so has required Canadian manufacturers to come closer to the levels of productivity of their foreign competitors. As a result, producers in Canada have adopted more efficient production techniques and management practices as well as becoming more outward-looking in their management philosophy and their approach to problem-solving. However, these adaptive responses have differed between the sectors of control and the indications are that firms in the Canadian sector of control have been more responsive than the subsidiaries. In other words, imports are becoming an increasingly more important source of technical change relative to direct foreign investment since the mid-1970s.

For instance, as has been discussed above, the subsidiaries were more specialised (both horizontally and vertically) in the early 1970s than their Canadian-controlled counterparts. Since then the Canadian-controlled firms have increased their degree of product specialisation and are now as horizontally specialised as the subsidiaries (MacCharles, 1984). There is also evidence that firms in the Canadian sector of control increased their degree of production specialisation relative to the subsidiaries by increasing their contracting-out of components and services to suppliers. Consequently, by the late 1970s the difference in vertical integration between the sectors of control had narrowed considerably. Further, in contrast to the Canadian-controlled firms, many subsidiaries had actually increased the diversity of their operations through increased contracting-in of production. This increase in vertical integration is based upon observed decreases in the ratio of purchased material to value added for the manufacturing sector as a whole as well as on interviews with a sample of firms (Daly and MacCharles, 1986b; MacCharles, 1984).

These differential responses between the Canadian and foreign sectors of control probably improved the productivity and competitiveness of the Canadian sector and narrowed the productivity gap noted in Table 4.2 that existed between them in the early 1970s. The improved performance of the Canadian-controlled firms is probably related to the greater pressure on them from the increasing levels of competition (especially for the smaller ones) because of their relatively greater diversity and lower profitability in the early 1970s. These factors help to explain the improved export performance of

Canadian-controlled manufacturers over the 1970s, which now matches that of the subsidiaries (Baldwin and Gorecki, 1983a, b, c; MacCharles, 1984).

The differential responses and performance between the sectors of control was also noted in a study using a sample of smaller firms (Daly and MacCharles, 1986b). That study showed Canadian-controlled firms were more responsive than subsidiaries to the changing competitive environment, partly because the subsidiaries were relying unduly on firm-specific comparative advantages arising from their association with parents and affiliates. These advantages were acting as protection for the subsidiaries, slowing their response to the changing competitive environment. The most notable advantages of this association were access to the deep pockets and to the knowledge of parents in such areas as finance, marketing and R&D. In addition, other factors were at work such as: benign neglect by the parents of their relatively small Canadian operations until major problems occurred; and a greater degree of administrative rigidity within the large MNEs that slowed their response and dampened entrepreneurial initiative by Canadian managers. The successful firms in the sample acquired the knowledge necessary to be competitive by building firm-specific intangible assets through hiring top-quality people and spending on their training and development. They also relied on imported machinery as well as leading suppliers of components to provide them with the latest technical knowledge.

IMPACT OF MNEs ON RESEARCH AND DEVELOPMENT

The MNEs have been criticised for not doing enough R&D in Canada, with the Science Council of Canada a leading proponent of this view (Britton and Gilmour, 1978). Those holding to this view argue subsidiaries are truncated operations with inadequate resources devoted to R&D activities. This helps account for their high propensity to import, it is argued, and also makes them slow to innovate.

While R&D activity is a means for producing innovations which can be used by firms to improve products, management practices and processes, it is only one means for doing so (MacCharles, 1982, 1983). Subsidiaries undertaking more R&D in Canada may improve the rate of domestic production of innovations, but this will not ensure their diffusion and use by firms, which is the significant issue. The difficulties in Canada with productivity, costs and competitiveness do not appear to be attributable so much to a lack of spending on R&D by the subsidiaries as they are to the inward-looking orientation of managers in both the Canadian and foreign sectors of control. This orientation impedes the quick adoption of known technologies and management practices already in use in the rest of the world. The estimates of the National Research Council are that 98 per cent of world expenditures on R&D are outside Canada (Palda, 1984). This suggests more emphasis needs to be put on diffusing existing innovations sooner rather than focusing on their production. Further, for reasons already discussed, subsidiaries are at least as productive as Canadian-controlled firms. The lack of innovation and diffusion appears to be related mostly to the smaller Canadian-controlled firms than to intra-firm sourcing by subsidiaries.

There are some differences in R&D activities between the sectors of control. The subsidiaries largely leave basic R&D to parents and affiliates. They stress adapting the products of affiliates, as well as innovations in production methods, while product innovations are preferred by Canadian-controlled firms (DeMelto, 1980). This may account, in part, for the higher value added per worker noted in Table 4.2 for the smaller subsidiaries.

Leaving basic R&D to parents and affiliates represents desirable specialisation from the standpoint of the MNEs. Centralisation of R&D activities in larger parents allows MNEs to benefit from economies of scale in its production. Parents and affiliates can undertake basic R&D more efficiently than smaller subsidiaries in Canada because of their larger scale of operation which allows them to more effectively utilise any indivisible inputs required in its production. The results of centralised R&D are freely available to the subsidiaries since it is profitable for MNEs fully to utilise firm-level knowledge and thereby maximise their rents. At the same time, subsidiaries efficiently undertake smaller scale applied research projects that are close to the customers utilising them.

Theory and evidence suggest, therefore, that Canadian subsidiaries do have access to the knowledge produced by the R&D efforts of parents and affiliates and that it is both high in quality and available at minimum cost and often priced out to the subsidiaries at a zero value. Any 'truncation' of R&D activities in subsidiaries is related to the ability of MNEs to engage internally in the international specialisation of R&D resources. This behaviour does not appear to be detrimental to the productivity performance of subsidiaries.

ORGANISATIONS AND INTERNATIONAL KNOWLEDGE TRANSFER

The importance of organisations to the process of diffusing technological change should not be overlooked. For instance, it takes good management practices for firms successfully to bridge the gap between the supply side of technological advance and its application to products and production processes that are in demand. Also, resistance to the faster adoption of known technologies is often related to the organisational environment in which they must be diffused and this involves both the social and managerial values in which the organisation functions. It is critical to have research done in Canada on why some organisations, managers and workers impede innovation and the diffusion of knowledge. While other countries will continue to engage in scientific R&D and the production of technological knowledge from the theoretical and applied hard sciences, no other country is going to do the research for Canada in the social science and managerial areas on why these new technologies are adopted more slowly here.

Management attitudes are a significant part of the problem of poor international transmission of technology and knowledge in Canada. There generally is a lack of awareness by managers in both sectors of control of how significantly the environment around them has changed and then how to adapt to it (Daly and MacCharles, 1986b). The need for increased specialisation, exporting and use of foreign suppliers involves a redirection of corporate philosophy, strategy, and organisation from that which served Canadian companies in a more protected era. The costs of changing corporate strategies and the intellectual capital needed to cope with new strategies can be considerable.

Managers in subsidiaries have been slower recently to adapt than have managers in Canadian-controlled firms because of factors in the former's operating environment. It is not easy to change the mission of a firm from an inward-looking import competer to an outward-looking exporter. It requires the active participation of corporate-level managers as well as local ones and the development of strategies for their subsidiaries which are consonant with corporate objectives on a North American basis (Daly and MacCharles, 1986b). Autonomous Canadian-controlled firms do not have to deal with the same depth of bureaucracy so they are able to adapt more quickly. They also have to adapt quickly since they do not have the same access to financial resources as do the subsidiaries, access that allows a delayed response.

There is general agreement that manufacturers have the ability in a technical sense of being able to specialise production systems and this would be fostered by freer trade and increased competition. Manufacturers have already been adapting as competition increased over the 1970s and into the 1980s, although the relative steepness of the Canadian recession in 1982 suggests that more firms should have been rationalising sooner. It is also generally agreed that the adjustment process and reallocation of resources in Canada accompanying such rationalisation could be relatively inexpensive and smooth. One estimate is that the cost would be roughly 6 per cent of the total value of the resources being reallocated (Harris, 1983). The adjustment costs are low because rationalisation involves intra-firm rather than inter-industry and inter-region reallocation of resources.

However, these estimates for the cost of rationalisation may be too low. They do not take into account the difficult and costly process of reorientating the planning and strategic decision-making processes in firms so the managers can understand and cope with their changing environment. Models of adjustment that use data based on existing structures to estimate the costs of adjustment would not likely be accurate because they cannot capture these organisational rigidities and their impact on economic activity. Freer trade and increased competition, in conjunction with improvements in the level of management education and training in Canada, would go a long way in overcoming inward-looking management attitudes which retard the diffusion of innovation quickly in Canada.

It should be understood that most of the productivity gains associated with improved technology and knowledge will have to go to reduce prices. In the average case, little will be left over to improve wages and returns to capital since price reductions are needed to achieve competitiveness in world markets of about the same magnitude as the potential productivity improvements from specialisation. Unless the prices of manufactured goods are reduced to competitive levels then firms will not be able to pursue policies of increased specialisation through increased exports. Further, continued productivity improvement beyond this will be needed since competitors elsewhere in the world will continue to move ahead.

Unfortunately, the Canadian experience over the past decade, in which real wages increased more quickly than productivity, has resulted in an increasing share of output going to wage increases at the expense of capital and market share (Daly and MacCharles,

1986a). It will be difficult to ask labour to retrain and accept automation without the incentive of increased earnings to do so. The experience of the 1981–2 recession was that US workers were more willing to accept wage cuts in order to restore the financial health of their employers than were Canadian workers. Moreover, organised labour in Canada often views technical change as job-destroying rather than job-preserving. Clearly, the co-operation of labour on technical change is necessary as are measures to assist with labour's adaptation to it. Improved labour-management relations and the organisational structure in which the change has to be diffused are keys to the success of technological improvement in the workplace. A shift in policy emphasis is needed toward greater use of relocation and retraining assistance for labour, in contrast to the much more expensive, current policies which stress keeping workers in obsolete jobs at the expense of resource reallocation and the improvement of productivity (Jenkins, 1985).

CONCLUSION

The international transfer of knowledge is effected in many ways as part of the normal economic activities in a society. The major international transmission mechanisms (assuming immobility of labour) are trade, direct investment and licences. The more competitive the economic system then the better the rate of innovation and diffusion of new knowledge is likely to be.

The objective in transferring and diffusing technology is to improve productivity and lower unit costs so that society can be on a higher production possibility frontier with full employment and an efficient allocation of resources. Neither of these objectives are being achieved at the present time in Canada's manufacturing sector. Therefore, the issue of productivity improvement generally, along with the related and more specific issue of the international transfer of technology and knowledge, remain important topics for public discussion and policy in Canada. However, the importance of international knowledge transfer should not be lost in the side issue of the nationality of those who are doing the transferring to Canada.

Part Two: Multinational Enterprise Strategies and National Policies

5

Contractual Agreements and International Technology Transfers: the Empirical Studies

Bernard Bonin[1]

In 1958, J.N. Behrman[2] published a monograph on the licensing experience of US firms. To our knowledge, this is the first empirical study on contractual agreements as a channel for international transfers of technology. In this study, the experience of 207 US firms was analysed and some 4,000 licensing agreements were investigated, about two-thirds of which were between independent firms and one third between affiliated companies. Valuable information can be found on the geographical distribution of the agreements, their coverage and structural characteristics, the types of payments involved in both cases, and the typical royalty rate.

The aim of our paper is to show how the analysis of international licensing has evolved since Behrman's pioneering work. The focus will be on the empirical studies of the behaviour of direct parties to the agreements (i.e. sellers and buyers of technology) in the context of an imperfect international market for technology.[3]

THE SELLER'S OPTIONS: INTERNALISATION VERSUS CONTRACTS

If a firm is free to choose between internalising its firm-specific advantage and transferring it through licensing agreements, the latter will seldom be chosen (Wilson, 1975; Davies, 1977; Telesio, 1979).[4] In Wilson's work, contractual agreements are a strategic decision of the firm which is part of oligopolistic rivalry. Oligopolists will try to segment markets in which they sell; they will also tend to engage in product rivalry. Contractual agreements, and more precisely licensing, will play an important role in market segmentation as well as in product rivalry. The seller can segment markets by defining geographical or functional territories in which he will allow his invention to be used; for instance, by allowing the buyer of technology to sell only in a predetermined market or to use it strictly for the production of some specified products. Licences can also be one element of product rivalry, since one form this rivalry takes — the one based on the physical attributes of products — is a function of R&D outlays. Although licences can also be an element of oligopolistic co-operation, with firms avoiding some of the costs inherent in product-rivalry, in many cases this rivalry will seem so attractive that contractual agreements will be considered as being too limiting.

If this hypothesis is well founded, there will be clear differences between licences granted by US firms for production in the United States and those that concern production abroad, the latter being more frequently and actively sought after than the former. The hypothesis appears to hold based on Wilson's results. He believes this is because firms try to keep their markets geographically segmented by not granting licences to competitors for production in their national market, while international licensing agreements become relatively attractive given the presence of barriers to entry erected by oligopolists abroad. However, more frequent territorial restrictions can be expected in international licensing agreements than in those dealing with production in the national market of the licensor.

Davies' results point in the same direction. His study, based on data collected by questionnaires and interviews, concerns the experience of British firms in India. The number of contractual agreements increased at the end of the 1960s following Indian protectionism. However, only firms for which the Indian market represents a substantial proportion of their total sales have tended to opt for an alternative to internationalisation in response to Indian protectionism. Clearly, licensing agreements were perceived by firms as a second-best solution, dictated only by market or legislative constraints.

Telesio's analysis shows that small and medium-sized firms tend to use licensing more than larger ones. Licensing agreements are seen as a way to penetrate markets abroad, where the oligopolistic competition more or less compels small and medium-sized producers to follow larger ones. Licensing agreements are again seen as a second-best solution by small and medium-sized licensors; however, because of their limited resources, they will be frequently used. Often, licences will be treated as a first step, leading eventually to participation in the licensee's equity. Frequently, contracts will include such an option at a future date in favour of the licensor; moreover, 40 per cent of the small and medium-sized licensors see the licensee as their eventual partner in a joint venture, compared to only 9 per cent for the larger ones.

However, if the owners of technology are not very enthusiastic about contractual agreements, the preferred alternative — that is, internationalisation through foreign direct investment — should be gaining ground when firms are free to choose between the various possibilities. Indeed this is what another empirical study shows (Davidson and Harrigan, 1977). From the evidence on 44 US firms which introduced 733 new products on the market between 1945 and 1976, it is clear that innovations are very often introduced subsequently in foreign markets where cultural affinities with the United States exist. However, in view of a substantial increase in the absolute number of products introduced in foreign markets in the 1960s, the combined share of Canada and Great Britain decreases from 40 per cent in 1945–55 to 28.8 per cent in 1955–65 and to 16.4 per cent in 1965–76. Managers of multinational enterprises (henceforth MNEs) have in recent years been marketing their new products abroad, after first introducing them in the USA, sooner than in any period before. The proportion of innovations introduced abroad less than a year after being first introduced in the USA jumps from 5.6 per cent in 1945–50 to 38.7 per cent in the 1971–75 period. Between 1945 and 1950, innovations were transferred abroad through licensing and foreign direct investment in about equal proportions; later, internationalisation was used much more frequently than licensing agreements, although the trend appears to peak in the 1970s.[5]

But if internalisation is the preferred option, the question remains: why does a firm try to internalise as much as possible its technological advantage?[6] MNEs emerge because arm's-length markets for intangible assets are failure-prone, and the empirical evidence shows a strong presence of such firms in research-intensive industries. Foreign direct investment is normally preferred since the owner of the technology is thus in a position to capture all the rents attached to his technological advantage, while licensing is more risky in this regard. Contractual agreements will be entered into only when the potential benefit from intangible assets cannot be otherwise exploited.

In what conditions will owners of technology benefit from internalising transactions that involve a transfer of technology? Using data on 1,376 internal and external transactions to which 32 American MNEs were parties between 1945 and 1975, Davidson and McFetridge (1984) have attempted to identify these conditions. An external transaction is one in which equity participation between the firms involved is not higher than 5 per cent and an internal transaction one in which such participation is over 95 per cent. Their results indicate that internalisation will be favoured, for a particular host country, when a large fraction of preceding transfers were internal ones. Hence the type of transfer selected stems from characteristics of the host country. The probability of an internal transfer is stronger if there is already an affiliate in the host country, a proxy for experience of operations abroad; it may be noted that Telesio (1979) maintains that a lack of experience of such operations will favour licensing between independent firms, which seems broadly consistent with the results obtained by these authors. If the technology transferred is not only new but radical, Davidson and McFetridge show that internalisation will probably be chosen, although the age of technology does not seem to be as significant as the type of the preceding transfers. A firm in an R&D-intensive industry transferring its core technology or a major product will be likely to opt for an internal transfer. The higher the number of previous transfers done by a licensor through licensing agreements, the weaker is the probability that the next transfer will be an internal one.

The rationale for contractual agreements

Although it is failure-prone, the market for technology exists, and firms use contractual agreements to some extent. Why is this so?[7] In certain conditions, specific to systems, to the firm, the industry, the host country or the country of origin, licensing will be advantageous to the owners of technology.

The advantages of licensing would depend on: (a) the characteristics of the technology involved (licenses will rarely be used if the transfer involves a core technology of the licensor rather than a peripheral one; they will be more frequent for old technologies than for newer ones, except if the pace of technological change is sufficiently fast so that the leader can stay ahead even if he cannot stop competitors from copying it); (b) the size of the firm (small firms will tend to use licensing more than larger ones, since they lack the necessary resources for foreign direct investment); (c) the maturity of the product (licenses will be more willingly granted for relatively old products, except if technological feedback or reciprocity looks good even for newer products); (d) the firm's degree of experience in international operations (risk considerations; comparative pace of response for licensing and foreign direct investment; transaction costs relative to licensing); (e) constraints related to host countries and to countries of origin (barriers to entry of foreign direct investors; an opportunity cost of capital which is higher in the host country than in the country of the potential licenser will be detrimental to licensing since the licensee would thus put a lower value on the flow of rents expected from the technology than would the owner of the technology himself).

Given these advantages, one would expect licensing, and more generally new forms of international investment, to become increasingly important (Telesio, 1979; OECD, 1984). Besides licensing, the latter would include franchising, management contracts, turnkey operations, co-production agreements, international contracting-out and joint equity ventures in which equity participation would be 50 per cent or less.[8]

However, Canada is an exception to the rule (Davidson, 1981). This country tends to get American products earlier than other countries, albeit various elements seem to be altering that trend. But only 5 per cent of all transfers of technology to Canada are done through licensing agreements and close to 90 per cent of transfers go through wholly-owned subsidiaries of US multinationals. A similar situation exists in no other country in the world.

Characteristics of transfers through different channels

It can be expected that the technology transferred through internalisation will be to some extent different from that more willingly transferred by contractual agreements. First, with regards to the type of technology transferred,[9] a subsidiary is more extensively used to transfer a new technology (i.e. for the first five years following its introduction). A new product will be more often transferred to a subsidiary than a new process which is more frequently transferred through simple export of machines embodying the process. Innovations with a rather low profitability will be transferred through contractual agreements, and those offering a higher profitability through internalisation. Licences become more frequent after five years. The age of technology transferred to subsidiaries in developed countries is six years on the average, while it is ten years for that which is being transferred to less developed countries. The transfer effected through licensing or joint ventures involves a technology that is older still, averaging 13 years. Subsidiaries in developed countries have recently been getting newer technologies; this trend is not apparent for technologies transferred to subsidiaries in LDCs or for transfers through licensing agreements. Crookell (1984) maintains that licences will be rare for a core technology, except in the case of an old technology widely available amongst competitors and also in the presence of a cross-licensing agreement with another R&D-intensive firm. A peripheral technology will be more frequently transferred through licensing if it has an uncertain commercial value or if the owner has opted to withdraw from that particular activity.

Davies (1977) shows that the quality and extent of assistance included in a technology transfer project is related to its expected profitability and to the owner's ability to capture such profits. The quality of assistance and the availability of transferred resources tend to be better in joint ventures than in licensing, based on his study of the experience of UK firms in India. Telesio (1979) shows it is difficult to get a licence from a large firm whose operations are little diversified; his evidence also shows that it is difficult to get licences in the pharmaceutical, chemical, electrical and electronic industries if no technology is offered in return.

There is another aspect to this question of the degree of availability of a technology — will the transfer be as complete in a contractual agreement as in internalisation? We know, from Davies' work on UK firms in India, that a participation in equity tended to lead to a fuller transfer than simple licensing; in other words, greater efforts were made to arrive at a successful transfer of technology in such cases. The dominant attitude of US firms has not been very different. On the contrary, they have never hidden their preference for the wholly-owned subsidiary. Again recently, Behrman and Wallender (1976) have done extensive, in-depth interviews with MNEs and come to the conclusion that 'only with the closest of business associations (a wholly-owned relationship) can an affiliate expect to tap fully the technology of the parent company' (p. 19). Put differently, since foreign direct investment is deemed preferable to contractual agreements, from the standpoint of the owner of a technology who is attempting to keep all the rents tied to monopolistic advantages, one can expect fuller transfers in the first type of relationship than in the second. In a relationship which would not involve substantial participation in the buyer's equity, such as in pure licensing, sellers will likely bargain for fixed payments which are as high as possible, in exchange for as little assistance as possible.

The presence of restrictive clauses, imposed by sellers of technology, is also notable in agreements. Indeed, there is also market sharing within MNEs, so that a particular subsidiary might not be authorised to export to such and such foreign market. But such restrictions are also quite frequent in joint ventures and other contractual agreements. The owner of the technology is thus trying to regulate the competition he will have to face. From the evidence based on 257 agreements involving 22 firms, Crookell (1984) finds that restrictions imposed by non-diversified firms are much more frequent. For instance, notwithstanding patent transfers, a grant-back clause is present in 99 per cent of agreements involving non-diversified licensors, by comparison with only 34 per cent for diversified firms. On the other hand, a clause of territorial exclusivity is more willingly granted to the licensee by non-diversified than diversified firms. Export restrictions are quite frequent.[10]

The question of the suitability of technology to host countries' conditions has also been the object of some empirical work, especially in relations with less developed countries. First, there is the question as to the existence or not of alternative technologies. All industries do not offer the same number of options in this respect: industries based on mechanical processes often rely on standardised equipment, easily available off-the-self; in others, such as those based on chemical processes, equipment is very specific to each plant, and alternative technologies are not often available.

Second, questions arise as to the criteria used in selecting a technology. Contrary to what proponents of appropriate or intermediate technologies often claim, the trade-off between capital and labour cannot be the sole criterion: factors relative to management must also be considered. More automated methods of production may then be chosen because they allow for better quality control; the minimisation of wastage or losses; quicker response to demand fluctuations; the reduction of manpower training costs; the alleviation of labour relations problems; or simply the prestige of relying on the most up-to-date equipment.[11] In general, there is little adaptation of the technology that is being transferred abroad. However, when such efforts are made, they tend to be more frequent in joint ventures than in simple licensing and, when limited to licensing, more frequent in transfers to LDCs than developed countries (Davies, 1979; Contractor, 1981).

Lastly, fixing the price of a licence can be a very complex

decision since the market is very imperfect and parties to the negotiation not very well informed (Caves *et al.*, 1983; Contractor, 1981). There is no standard model of price determination. The price will be likely to fall between a minimum set by the costs incurred to make the transfer effective and a maximum which will reflect the value of the technology to the buyer or a limit imposed by policies of the host country. In part of his analysis, Contractor takes account only of agreements between independent firms, since payments related to licensing agreements between parent companies and affiliates may be affected by fiscal considerations (Kopits, 1976; Mathewson and Quirin, 1979; Rugman and Eden, 1985).

On the characteristics of the technology transferred through various channels, Behrman and Wallender (1976), Contractor (1981) and Teece (1976) have obtained results that are worth summarising:

(a) It is not always possible to delineate precisely the technology that is being transferred; consequently, there is no easy way to determine the costs of the assistance involved. Methods of compensation for the technology will vary from one firm to the other but, at best, the relation between payments and the costs involved will be very loose, especially when a transfer to an affiliate concerns a vast array of techniques, processes, specifications and know-how. Cost allocations will necessarily be arbitrary or the result of an agreement between free parties; firms will have some room to manoeuvre with regards to internal transfer pricing.

(b) A successful technology transfer involves costs for the MNE; hence the hypothesis of a zero cost operation can safely be discarded. However, payments obtained in such operations can be a multiple of the costs incurred in a transfer, especially when it goes to a producer that is not affiliated to the owner of the technology.

(c) Many transfers concern only an already developed technology, having passed the test of time, so that relatively little applied research will underlie the immediate transfer operation. The most difficult transfer — and hence the most costly — is that of a previously little tried technology, of a technology that is available from very few competitors, or that is being transferred shortly after having been tried for the first time. Then royalties can clearly be very high, particularly if there are few alternative sources for the technology.

(d) Teece's results tend to confirm the hypothesis to the effect that an internal transfer to a wholly-owned subsidiary is less costly than any other form of transfer, the magnitude of some of the cost differentials involved being rather impressive. On the average, a transfer to a joint venture costs 5 per cent more than to a wholly-owned subsidiary while it is 9 per cent more for a transfer to non-governmental enterprises in which there is no equity participation by the licensor, and 17 per cent more for transfers to state-owned enterprises with no equity participation.

(e) The costs of transfers seem to decrease after the first transfer and with greater diffusion of the technology. Technology transfer would thus appear to be a decreasing cost activity, which might explain why some engineering firms would specialise in turnkey operations. It is then possible to get higher royalties (expected profitability) in subsequent operations.

(f) By comparison with other possible channels of diffusion, the MNE seems to be particularly well suited to the grouping of the ingredients of a successful transfer. Not only is it able to transfer technology at a lesser cost, but an internal transfer will probably be more complete than by any other channel, although some constraints on the licensee will be likely to exist. An MNE will not always transfer a technology well adapted to the relative factor prices of a host country (a question raised mainly by LDCs), but it does not seem to be more reluctant to adapt the technology than when the transfer is done otherwise. Yet, once the international transfer has been effected, the MNE does not clearly offer the assurance of a more rapid diffusion of the technology inside the host country.

CONTRACTUAL AGREEMENTS FROM THE BUYER'S STANDPOINT

Killing (1975, 1980)[12] has analysed the conditions in which licensing can be a viable growth strategy for Canadian firms, and a weak R&D effort can be compensated for by the acquisition of licences. His analysis deals with transactions between independent firms, and is based on a model of the conditions in which licences are traded: R&D expertise of the buyer; single transfer versus a durable and updated relation; presence or not of restrictive clauses. Buyers of technology, strongly involved in R&D, face no particular constraint, no obstacle to their growth. However, they learn very little from the licensor and do not get access to new areas of growth, if the agreement is for a single transfer. Moreover, such agreements for the transfer of know-how and patents frequently include restrictive clauses, although the buyer might be getting access to new information and new sectors to develop; but because of his own expertise in R&D, growth remains possible.

On the other hand, licensing, even for a continuous transfer, is not a viable strategy for firms with no particular expertise in R&D, except maybe for products which come late in the product cycle. Not only will they face important restrictions, but, if they opt for a very specialised market niche, they will operate on very limited markets, and if they try to compete with subsidiaries of MNEs, their task will be made difficult because of their weak R&D effort. Unless such transfer agreements are seen as a way to build their own R&D expertise, as is sometimes the case, a licensing strategy is not a viable one.

In his 1980 article, Killing uses data on 74 licensing agreements and 28 joint ventures to study the determinants of the type of transfer: licence; licence with a continuous updating of the technology transferred; joint venture in which the buyer is the majority shareholder (70 per cent or more of the equity); and joint venture in which the buyer does not have control (50 to 55 per cent of the equity). He hypothesises that the more a buyer is in need of knowledge to use a specific technology, the tighter will be the relationship between seller and buyer. The degree of diversification that the buyer hopes to achieve will determine the extent of the necessary learning and, hence, the length of the relationship for which the buyer will aim. The type of transfer will stem from the objectives of the buyer, for the length and intensity of the relationship tend to increase from the simple licence agreement to a joint venture with no control by the buyer. If firms do not systematically go for the closest possible relationship, it is because the cost of such a relationship (royalties plus restrictive export clauses, for instance) increases with the closeness of the relationship.

If a subsidiary is seen by a host country as a form of dependant, it is very unlikely that contractual agreements will be better in this respect. When a buyer tries to get a valuable technology through licensing, there will probably be constraints imposed on his decision-making autonomy, unless he is able not only to absorb the technology acquired but to improve it. Restrictive clauses are a reflection of market imperfections confronting the buyer. Since the licensor is unable to keep all the rents for himself, he fears that the licensee might eventually become a competitor; consequently, he imposes various restrictions to limit the competition.

CONCLUSION

(a) Since Behrman's pioneering monograph of nearly three decades ago, analysts, with few exceptions, have tended to move away from the analysis of licensing agreements as such. Instead, most recent studies see contractual agreements as an alternative to internalisation in the main, and try to identify the determinants of the various options.

(b) The market for technology is admittedly imperfect, but it exists nonetheless. At the outset, the position of the buyer would appear to be weak; however, over the years, it tends to get stronger.

(c) From the seller's standpoint, licensing is clearly seen as a second-best solution. Generally, he will prefer to internalise his technological advantage, should the host country leave such an option open to him. For the buyer as well as for the host country, a transfer through a contractual agreement will often seem preferable to the presence of subsidiaries, since they are hoping to separate the package offered by the MNE.

(d) Licences are not clearly gaining ground compared to foreign direct investment; there is no clear consensus among analysts.

(e) There seems to be a relation between the mode of transfer and the characteristics of the technology transferred, but the picture becomes blurred when one tries to identify precisely the type of relationship involved.

(f) The contractual agreements between parent companies and affiliates have not received enough attention. For instance, when Teece (1976) observes that there are substantial cost differentials between an internal transfer and other types, the question still remains: is an internal transfer really less costly, or is it rather because the affiliate will not necessarily be billed for all the elements of the cost by the parent company, while efforts will be made to fully bill the licensee when a licensor has no participation?

NOTES

1. The author wishes to acknowledge Mr Roger Verreault's valuable assistance in preparing this paper. However, any error of fact or interpretation is the author's sole responsibility.

2. See Behrman (1958, 1960). See also the *Conference Supplement* published by the Patent Foundation of the George Washington University in 1959.

3. See Caves, Crookell and Killing (1983) and Caves (1983a). In other words, even if arm's-length markets for technology are failure-prone, they nevertheless exist. Baranson (1978) maintains that the bargaining power of the buyer of technology has been getting stronger over the years. Buyers are becoming better informed and face more numerous sellers, so that US firms have revised their technological policies substituting the sale of technology and management services for foreign direct investment.

4. For a theoretical analysis of the choice between internalisation and contractual agreements, see Casson (1979).

5. Market structures can also impact on the choice between internalisation and contractual agreements. Tilton (1971) shows that, in the semiconductor industry, the ease of entry of new firms has led to the rapid diffusion of the new technoogy in the United States, often through licensing. By contrast, in Europe it was the new subsidiaries of MNEs, and in Japan the large, well-established firms, which were mainly responsible for the diffusion of this innovation.

6. See Dunning (1981), Casson (1979), Caves *et al.* (1983), Davidson and McFetridge (1984).

7. For further comment on this question see especially Wilson (1975), Contractor (1981), Robock and Simmonds (1983), Buckley and Davies (1981), Caves (1983), Crookell (1984) and Telesio (1979).

8. The OECD group of studies, all published in 1984, can be found in the Bibliography. See Oman, Ozawa, Franko, Flamm and Pelzman, Pollack and Riedel, Delapierre and Michalet, Onida, Parry, Blömstrom.

9. Mansfield and Romeo (1980), Mansfield, Romeo and Wagner (1979), Caves (1983), Crookell (1984), Caves *et al.* (1983), Davies (1977), Telesio (1979).

10. See Caves *et al.* (1983), and Ariga in Sagafi-Nejad, Maxon and Perlmutter (1981). Safarian (1966), believes the presence of restrictions in agreements might have more to do with abuse of the international patent system than with foreign ownership. One could expect the financially independent firm finding the terms of agreements both explicit and restrictive as to markets, more so than a subsidiary in which gains inadequately spelled out can be recouped by profits or dividends transferred to the parent. Yet, wholly-owned subsidiaries with no export franchise dominated those with such a franchise by a two to one margin among his respondents in Canada. See pp. 141–2.

11. See Robinson (1979), Amsalem (1982), Contractor and Sagafi-Nejad (1981b).

12. See also, Crookell (1973, 1984), and Caves *et al.* (1983).

6

Multinational Enterprises: Transfer Partners and Transfer Policies

Gilles Y. Bertin

Asserting that technology transfer is of vital necessity to multinational enterprises (MNEs) sounds by no means original. But whereas much attention has been paid so far to the technology transfer from the MNEs outwards, a look at the Bibliography to this volume will confirm that relatively little exists on transfers inwards to the MNEs or on the global structure of their transfers. Using original data from a study on the patenting activities of MNEs,[1] this article aims to give some new statistical information on the structure of transfer to the various group of partners. It also provides some insights as to the relative importance of internal and external partners.

After a brief summary of what technological transfer means to MNEs, we consider the specific role played by MNE partners in transfer operations. We then present the statistical base and methodology used, and the main findings, before coming to some conclusions and questions.

TECHNOLOGY TRANSFER AND MNEs

The technological assets of a firm are one of the main, if not dominant, sources of its competitive power. These technical assets are the amounts of accumulated new inventions, original products, technical processes and experience, whether proprietary or not, which it utilises. They are valorised in productive activities through direct investment or indirectly through their sale or licensing. Transfer 'outward' may become necessary: (1) to valorise original techniques which cannot be directly used by the firm, (2) to provide additional help in conquering new markets both at home and in foreign countries, (3) to help in setting good technical relations or in sharing the cost of research on new products with competitors. But technique depreciates over time; it also has to be continuously updated and renewed, all the more when the firm, as is the case with MNEs, is engaged in worldwide competition. In large companies with developed R&D services much of the new technique is likely to originate from the firm itself. But in addition to the continuous free flow of fundamental scientific information, new ideas and techniques also have to be taken from competitors' or other firms' inventions and innovations.

This double flow of technique inwards and outwards has somehow to be balanced in order to preserve the technical power and independence of the firm. This means, first, that the firm should strive for the best possible valorisation of its technology to compensate as much as it can for the cost of fully self-produced and acquired technology. It also means it will have to get a close look at the partner to and from whom it transfers the technology. For MNEs there are at least two pitfalls to avoid: one is the transfer of advanced or sensitive technology to partners who might turn into dangerous competitors in the future; the second is the acquisition of such technology from a partner who may gain some type of control of a significant share of the firm's activity through its use. Both of these are threats to the firm's future power. To what extent can this double objective of securing a permanent flow of additional technologies and preserving the technical and financial independence of the firm be met? The question is largely left unanswered.

A second question arises from the physical and financial limits which the MNE, whatever its size, faces in its research activities. R&D is a costly part of business, the profitability of which can only be judged in the medium or long run. Due to the constraint of cost and to the export of specialised lines reflecting the firm's relative commercial advantage, MNEs are more likely to centre their own R&D on the main present or future lines of business. In these lines the largest part of the technology should be internally generated and utilised through direct investment. Transfer from and to partners external to the MNE should accordingly be scarce. But this policy would mean that more transfers would be needed from and to those lines of activity which the firm does not concentrate upon.

We would thus propose two possible models of transfer behaviour. In the first model, the firm holds a competitive position which is strongly centred in a main field of activity. It allocates the major share of its R&D expenses to this main field. It restricts itself to few transfers of technique, mostly with internal partners — subsidiaries or associates, whether foreign or domestic. In the second type, closer to what we know as the industrial conglomerate model, the firm holds no such definite and strong competitive position or it has several distinct ones. Accordingly its R&D is not as specialised and it turns to external as well as internal partners for frequent technical transfers to complement its own research activity or to valorise its own technical output. MNEs may belong to either type; more frequently, they are of the 'mixed' type, involving a principal activity comprising one or several major lines of business along with some secondary lines.

The assumption can reasonably be made, therefore, that the more concentrated and competitive the MNE is in a given field of activity, the more likely are internal partners to play a dominant role in the transfer of technique to and from the company, the less frequently will the firm apply for transfers to other external partners. In other words, an MNE with a large array of transfer partners should either be a large, diversified firm or a smaller, technically-dominated one. In any case, the choice of the potential transfer partners and their respective shares of total transfer are essential to the firm's technological strategy.

ROLE OF MNEs' TRANSFER PARTNERS

Current transfer partners belong to one of the following five groups:

(a) partners within the MNE, whether they are subsidiaries or simply associated firms,
(b) other large, independent MNEs,
(c) non-multinational firms which can be located either in the MNE's country of origin or in other countries,
(d) goverment or public firms or institutions at the national or regional level,
(e) university research laboratories or independent private research firms.

One can expect the average MNE to keep in touch with technical partners belonging to any of the five above groups or at least to be aware of the possibility of setting up such contacts should they be required. But transfers to or from any of these partners take place within different sets of conditions and do not offer the same mix of

Table 6.1: Reactions of transfer partners

Potential partners	Availability	Expected total cost	Expected total return
Internal partners	High	Low to medium	Medium to high
Other MNEs	Low to medium depending on market structure	Medium (through cross-licensing) to high	High
Smaller national firms	Medium to high depending on the technique and market structure	Medium to high	Medium to high
Public organisations	High	Low to medium	Medium to high
Academic, private research	Medium to high	Medium to high	Medium to high

expected gains and costs to the firm. Therefore, one may try to define the bases on which the choice of any given partner is made.

Three major criteria[2] by which the firm is likely to accept or reject transfer from or to any given partner are: (a) the conditions of *availability of the technique* which include the existence of a marketable technique and any special restrictive provisions for its use; (b) the *expected total cost of acquisition* (or *sale*) including the specific cost of adjusting the technique to this new environment and the so-called transaction costs; and (c) the *expected total return*, whether the technique is to be used inside the firm (for acquired technology) or outside by the firm or the acquiring partner.

Each transfer operation being a special case, no clear answer can be given on any particular occasion unless the real or expected figures can be estimated from past experience and the impact of the existing provisions duly accounted for. Nevertheless, the level of intensity of each criterion (high, medium or low) can be given on the basis of rational expectations from the past behaviour of each partner, its size, and its expected bargaining power. Thus general tentative conclusions could be drawn from the grid of criteria to show what the best probable transfer partners for the MNE could be, given other environment conditions such as: the speed at which technical change is occurring and will take place in the future; the prevailing competitive market structure; and the requirements of the technique such as the amount of investment needed, the time and scale of operation, and the size of the specific market relative to the total activity of the firm.

Table 6.1 indicates the rational reactions one may expect from each of the five above-mentioned types of partners. One may conclude from this table that MNEs will normally express a preference for internal partners. Not only is this choice the best from the viewpoint of both the availability of technique and the firm's independence but it also ensures lower (and flexible) costs.

In most cases, technical resources cannot completely meet the firm's demand for new technology. Additional resources should be sought from external sources, none of which can be regarded as entirely satisfactory. Availability ranges from low when the technique is under competing MNEs' control to high for government-owned techniques which can be acquired on easy terms by those companies ready for industrial production and sale. Total costs are always above the low costs incurred in internal transactions. They can even be high should the technique be acquired at arm's length from a unique MNE (or small company) owner and need some adjustment to new operating conditions.

Given these constraints, MNEs will probably take their techniques from diversified sources, depending mostly on the line of activity where it is needed and for which purpose. For instance, the probability is high that an MNE which is part of a worldwide oligopoly will acquire part of its technique from other rival MNEs through cross-licensing or joint research agreements. This is all the more the case when the cost of developing the new technique is high or short-term profitability is uncertain. Similarly, the MNE is more likely to transfer from and/or to smaller firms when and wherever it can only rely on quite limited resources of its own to conduct profitable research or investment.

What is actually the share of internal transfers in total transfer activity? To what extent does the reality verify the proposed choices of partners?

STATISTICAL BASE

To attempt to answer at least some of these questions we utilise the results of our questionnaire on the patenting and licensing policies of MNEs.[3]

Though the questionnaire did not specifically bear on transfer, but rather on patenting, at least one section of it (section III) dealt with transfer through patent licensing. In two separate but linked questions in this section, companies were asked to (or from) which other companies they granted (obtained) licences and by what percentage. Two more questions bore (1) on the main provisions included in the transfer contracts and (2) on the share of technology income derived from patent licensing.

In addition to this section, two questions (6 and 10) on the firms' general and foreign patenting policy explicitly refer to transfer policy as one of the main possibie motivations for patenting, or for patenting in foreign markets.[4] While answers to these two questions can help to establish the general framework in which licensing transfers take place, answers to the two other questions on the licences granted or obtained can provide some interesting quantified information. Not only do the latter disclose the companies' preferences for one specific type of partner compared to others but they also give substantial information on the very nature of licensing — for example, do firms prefer that it be open or internal, and do they prefer equal or smaller partners?

To get information suitable for present purposes we first retain only five types of partners; company's subsidiaries type (a), company's affiliates (b), other large MNEs (c), local companies in the home country (d), and local companies in other OECD countries (e). Three types of partners are omitted here, namely local companies in East European countries, local companies in LDCs, and other non-specified partners. These five types could be reduced to three or four if we consider that the company's subsidiaries and affiliates are both internal partners, and that there is little significant difference between local companies in home or other OECD countries (in both cases they are small, multinational external partners, but the domestic firm is closer to other national MNEs than are foreign firms). Further, to determine what the dominant transfer strategies are, we only take into account those types for which the share in transfers exceeds 20 per cent. This means that in most cases the number of 'significant transfer partners' is limited to two or three. This enables us to retain a limited number of strategies according to the partner (or the combination of partners) preferred in both outward and inward transfers. Such strategies may be defined as shown in Table 6.2.

The preferred transfer policies can be scaled from 1 to 8, which approximately covers situations from the most internal to the most external. If we then plot each firm's outward and inward groups of preferred transfer partners (or transfer preference) on the same graph, we may visualise the cloud of total transfer preferences

Table 6.2: Dominant transfer strategies

Strategy	Partner(s) preferred	Nature of strategy
1	a or b	Totally internalised transfer policy
2	a/b with d/e	Semi-internalised transfer policy
3	a/b with c	Mixed internalised and oligopoly transfer policy
4	a/b with c, d/e	Limited internalised transfer policy
5	a/b with c,d,e,	Open policy
6	c with d/e	External mixed transfer policy
7	d/e	External transfer policy
8	c	Pure oligopoly transfer policy

within one particular industry.

Let us assume, for instance, that the MNEs belonging to one given industry draw most of their acquired technology from external partners such as other MNEs or small foreign firms, and transfer their own technology outwards mainly to their own subsidiaries or affiliates. The cloud of dots will occur in the upper left quarter of the graph (Figure 6.1). Conversely, it would be located in the lower right quarter should the incoming technology be acquired from internal sources and the produced technology be transferred to external partners.

The methodology followed here raises some questions. First, one may question the validity and appropriateness of the data used. Results from a questionnaire on patenting behaviour are more likely to reflect transfer practices based on patent licensing rather than total licensing of both patents and know-how; the latter accounts for the larger share of total transfers. However, the evidence of such bias was not found as the author checked the validity of questionnaire results through interviews. This may be easily explained. Though pure patent licensing only accounts for about 20 per cent of total transfers, patents are also included in 'package transfers' along with know-how, trademarks, etc. — which make up to 60 per cent of total transfers. Respondents were largely unable to make clear-cut distinctions between pure and mixed licensing and gave overall answers. A second difficulty could arise from the fact that respondents to the questionnaire were mostly managers of industrial property, not directly involved in transfer policies in many companies. But this objection did not hold as transfer decisions were found to be taken usually by committees to which both industrial property and transfer managers belong.

The fact that about two-thirds of large companies with separate,

Figure 6.1: Transfer preferences for an industry

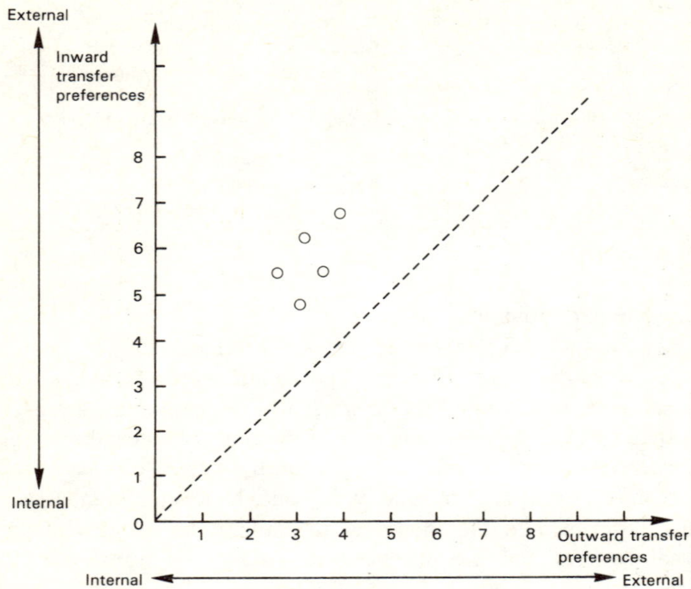

independent divisions gave only one overall answer could be more cumbersome. It is difficult to know to what extent this restriction alters the findings. Clearly some MNE divisions display independent transfer policies which reflect differences in their own technical and competitive environment. A few companies gave separate answers (one company gave up to four) but most did not; some of these apply common rules to transfer policies whatever the particular industry, but the lack of detailed information somewhat blurs the findings for a number of larger electrical or chemical MNEs. However, this may be partially offset by the fact that these companies checked a larger number of partners than did smaller firms, showing that, on the whole, they were engaged in mixed-open policies.

MAIN FINDINGS

The main findings of the inquiry may be summed up as follows. A total of 113 separate (complete or partial) answers on transfer from

Table 6.3: Number of responses on technology transfers[a]

Firms	Pharmaceutical	Other chemicals	Electrical and Electronics	Mechanical eng.	Automobile	Resource based	Total
US and Canada	9	16	14	9	4	3	55
European	10	6	11	5	5	12	49
Japanese	1	4	3	—	1	—	9
Total	20	26	28	14	10	15	113

Note: a. Additional information was derived from interviews including a host of firms, most of them European, which did not complete the questionnaire.

Table 6.4: Importance given to transfer as implied by patenting

	Pharmaceuticals	Other chemicals	Electrical	Mechanics	Automobile	R.B.	Total
Patent in view of future licensing agreements	5/19[a]	13/25	18/28	6/14	3/10	5/14	50/110
Patent in foreign countries in view of future licensing agreements	8/19	16/25	22/28	11/14	5/10	12/14	74/110

Note: a. Number of positive answers over total number of answers.

Table 6.5: Ranking by percentage of positive answers

Importance given to transfer as a motivation for patenting: % of answers		Importance given to transfer as a motivation for foreign patenting: % of answers	
Electronics	64	Resource based	86
Chemicals	52	Mechanical	79
Mechanical	43	Electronics	79
Resource based	36	Chemicals	64
Automobile	30	Automobile	50
Pharmaceuticals	26	Pharmaceuticals	42

95 companies were received. The number of answers by broad industry groups and country of origin of the MNE are shown in Table 6.3

The importance implicitly given to transfer policies can be derived from answers to questions 6 and 10 of the questionnaire. Patenting firms were assumed to show significant interest in transfers whenever they made the h (question 6) or b (question 10) choice. The results are given in Table 6.4.

Overall it appears that a little less than half of the firms (45 per cent) patent with a view to transfer through licensing; the proportion rises to two-thirds (67.3 per cent) for foreign patenting. This means that a majority of firms consider transfer, at least in its outward form, as essential to their strategy on technology. Ranking by the percentage of positive answers reveals significant differences among industries, as noted in Table 6.5.

The preference for given partners and the subsequent choice of strategies are illustrated for each industry in Figure 6.2. A few dominant features appear in the choice of transfer partners.

First, the number of partners outwards is always larger than it is inwards. Second, strategies are more open on the inward side than they are on the outward. With very few exceptions (only two in resource-based industries, and one each in chemicals and mechanical industries) firms seem to prefer to have more partners for the supply of new technology than on the demand side. However a large proportion of firms (46/87 or nearly 53 per cent) appears to keep a balance between identical sets of partners inwards and outwards. This is especially true of the electronics industry where more than two-thirds of firms have the same partners on sales and purchases of technology; the importance of cross-licensing is one possible explanation for this result, assuming a smaller number of suppliers of new technology, since a very significant part of it is then

Figure 6.2: Transfer strategies by industry

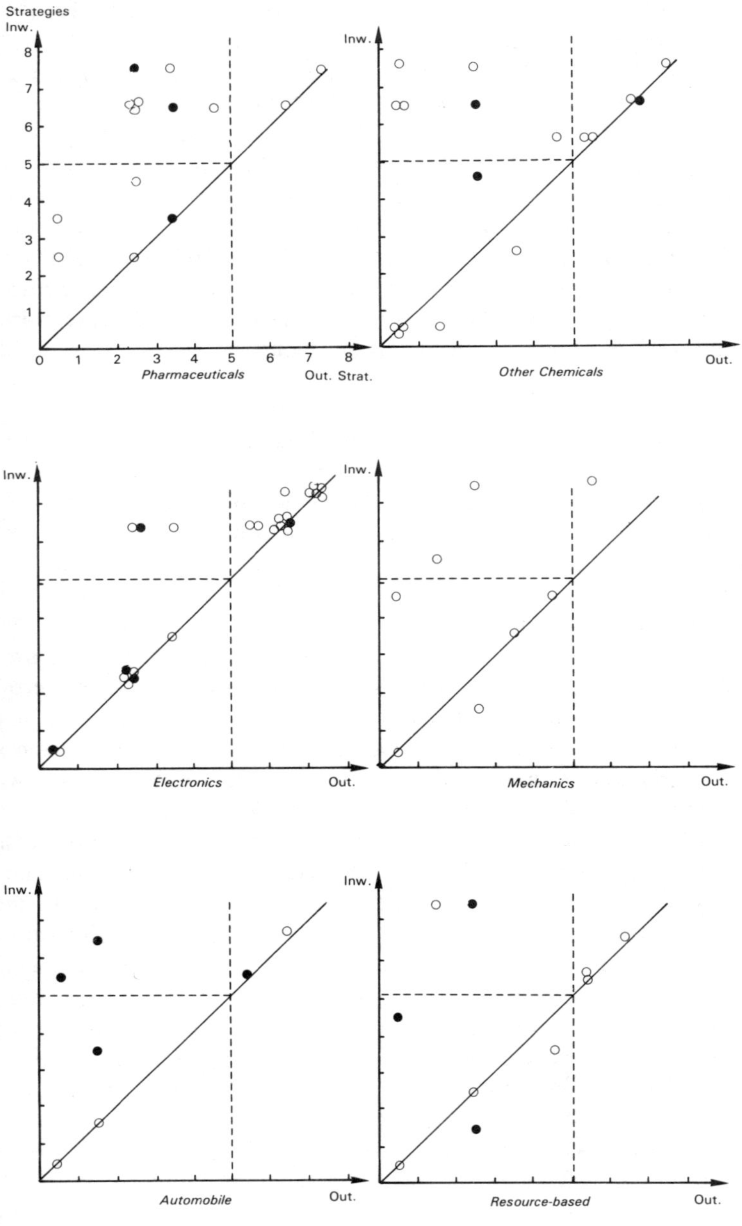

likely to be produced by the firm itself. The bigger is the firm, the more able it is to be its own supplier except in those fields, such as electronics, where the minimum size required for new R&D is large and compels the firm to be a partner of a global oligopolist.

The third main feature is the priority given by many firms to internal choices as opposed to external relations. With strategies 1 to 5 being considered as partly or totally internalised, about two-thirds of firms (64 per cent) have — at least to some degree — dominantly internalised links. Some distinction can be drawn between fully internalised firms (inwards and outwards) and partially internalised (outwards only). The second group of firms acquires a substantial share of its technology from external sources but largely transfers to its own subsidiaries and affiliates. Here again, significant differences appear among industries, as shown in Table 6.6.

Of the three distinct groups — fully, partially and non-internalised firms — the first two include the larger number of firms in all sectors except in electronics, where non-internalised firms slightly outnumber internalised firms. Pharmaceutical, chemical and mechanical industries are highly internalised, which can be explained by the high degree of specialisation and intense competition in world oligopolies. The automobile and resource-based industries show varied structures of transfer.

On the basis of the broad percentage range attributed by firms to each type of transfer, we tried to estimate what the mean share of total internalised transfer was for each industry. Of the 56 firms which quantified internalised outward transfers, 23 (41.1 per cent) reported shares between 1 and 20 per cent, 16 (28.6 per cent) between 21 and 40 per cent, seven (12.5 per cent), between 41 and 60 per cent, eight (14.3 per cent) between 61 and 80 per cent, and two (3.6 per cent) over 81 per cent. Thirty-one firms did not give any percentage at all or considered it to be too small for any range to be checked. The mean percentages by industry for reporting firms were as follows with the overall mean being 28.4 per cent:

Pharmaceutical	62.6%
Resource-based	35.0
Electronics	24.0
Mechanical Eng.	23.2
Chemicals	19.0
Automobile	8.6

Table 6.6 Internal versus external transfer strategies

Industry	Fully internalised	Outward internalised	Total internalised	Non-internalised	Total
Pharmaceutical	7	7	14	2	16
Chemical	6	7	13	5	18
Electronics	8	5	13	14	27
Mechanical	5	2	7	1	8
Automobile	3	2	5	3	8
Resource-based	5	2	7	3	10
Total	34 (39%)	25 (29%)	59 (68%)	28 (32%)	87 (100%)

These results can be compared with those of external transfer to other MNEs, both outwards and inwards:

	Outwards %	Inwards %
Electronics	42	47
Pharmaceutical	31	46
Chemicals	32	36
Resource-based	26	40
Automobile	23	29
Mechanical Engineering	14	16

Such results should be treated with caution. They probably underestimate the share of internal transfers of technology. Several reasons may be given for this, in addition to those mentioned above. In some firms internal transfers do not follow the same procedures and rules as those used for external partners, and may have been considered apart by respondents. On the other hand, transfers included in the operations of direct investment may be ignored, or not accounted for as technology transfers but rather as capital transfers. The answers given by a few US groups, which seem to deny the existence of any internal technology transfer, give some support to this hypothesis. Whatever the real figures are (and they probably run higher than our estimates), from 25 to 75 per cent of total transfers appear to be internal. The ranking by industry looks highly significant. Both the pharmaceutical and the automobile industry stand quite apart from other industries. Clearly these results are consistent with the importance of direct investment which is essential in the pharmaceutical industry; however, its role is relatively limited, as compared to export, in the automobile sector.

Differentiation by the size or the national origin of firms largely confirms anticipated results. The largest groups display more open contacts than smaller companies. This does not mean that the former's internal transfers are relatively less important, but secondary lines of activity within the main sector or outside of it are frequently more open, at least to outward transfers. This can be illustrated in the few cases where the companies gave separate answers by division. For instance, a large US company involved in diversified fields of electronics and computers follows a rather strict internal strategy based on transfers linked to direct investment in its main lines of business but shows external preference for the other two lines.

US firms do not show behaviour different from non-US (European and Japanese) groups except on the relative share of internal transfers, which is higher. This result is in accordance with the well-known US preference for fully-controlled direct investment.

CONCLUSIONS AND QUESTIONS

Conclusions to be drawn from our limited sample should be interpreted with caution, and need more testing by additional inquiries. As they stand, they tend to confirm existing views on the general strategy of large MNEs. They also largely corroborate the point that technology transfers should not be considered apart from other transfers such as through direct investments and exports, to which they are closely linked.

First, internal partners are preferred in most industries to external contractors, and among these latter, small domestic firms are preferred. This is a rather trivial result. MNEs are very sensitive to external threats and technology is a highly strategic matter, since an extensive diffusion of it could prove dangerous to the firm's future growth. As expected, external transfer partners play a significant role in industries of the world-oligopoly type, such as automobiles and electronics. What is rather new is that such partners are also significant in the chemical industry. The increasing cost of research may exercise a strong influence on the firm's strategy, as suggested in various interviews, compelling the firm to increase technology exchange with competitors.

Next, again except for world oligopoly sectors, internal partners

Table 6.7: Summary of multinational transfer strategies

		1	2	3	4	5	6	7	8
	8	2	1	3	1	—	1	1	6
	7	2	1	5	2	1	2	12	
Inward	6	3	2	—	—	1	5		
	5	2	—	1	—	1	—		
Strategies	4	1	1	—	3	—	—		
	3	1	6	7	1	1	—		
	2	—	1	2	—	—	—		
	1	10	1	—	—	—	—		
		1	2	3	4	5	6	7	8

Outward strategies

are given a strong preference in what could be called the core activities of MNEs. This is no surprise in a period of intense international competition. External partners are preferred in activities where they may offer new potentialities for growth in the near future (markets) or in the distant one (new research fields). It does not mean that the role of external partners is diminishing. Uncertainty and heavy costs of entry are grave concerns in new fields of activity such as new materials, communications and advanced computing. These concerns induce firms, even the largest ones, to look for extensive technical contacts and joint-ventures in research. In that respect, external partners, which are largely MNEs but also smaller firms from domestic and foreign countries, are likely to become full-time partners. The same is true for public and academic partners.

Last, is transfer easy to perform or not? This paper did not intend to give a direct answer to this tricky — and much discussed — point, but it casts some light on it indirectly. Usually transfer by itself is not difficult to internal partners who may have complete and free access to the common pool of technology and be given full assistance.[5] Nor is transfer difficult with equal external partners. This does not mean that the capital involved in such specific relationships is negligible but that the transaction costs are not so high as to deter the transfer practice. The trouble lies more with the economic environment in which the transfer takes place. Again there are few fundamental problems in transferring to other developed countries. But in LDCs even internal transfers to competent mature subsidiaries can prove inadequate or impossible when legislation, the economic situation or financial conditions make the transfer unprofitable or dangerous. A careful choice of transfer partners is then one of the prerequisites to any successful policy of technology valorisation.

NOTES

1. AREPIT (1985). The research was jointly conducted by AREPIT and the Science Policy Research Unit of Sussex University. The study was sponsored by the Institute for Research on Multinationals, Geneva.

2. We assume the existence of a previous demand for the given technique from the firm or from the external market.

3. In question 6, firms were asked as one of the 9 non-exclusive choices (choice b) whether they patented in order to have a patent portfolio with which to negotiate licensing agreements with other companies. In question 10, they were asked among 5 non-exclusive possibilities (choice b) whether they patented 'in foreign countries where you have entered or wish to enter into licensing agreements with other organisations operating within that foreign country'.

4. This is so with regard to the existence of links, but it does not mean that some are not stronger.

5. It can be when outward transfer is directed towards affiliates or subsidiaries in LDCs.

7

The Development of Technology in MNEs: a Cross-Country and Industry Study[1]

Hamid Etemad and Louise Séguin Dulude

During the past 25 years, a large number of theoretical and empirical studies have examined the relationship between technology and the multinational enterprise (MNE). These studies always implicitly assumed a degree of centralisation of research and development (R&D) and technological control at the MNE's corporate headquarters or in its home country. The literature on the advantages of centralising or decentralising R&D in MNEs suggests numerous factors to explain the phenomenon. Empirical studies have confirmed the high degree of centralisation of R&D at the MNE's corporate headquarters or home country. Some of these studies point to notable differences among MNEs.

Many factors are known to explain the subsidiaries' participation (or lack of) in the MNEs' R&D activity. They are the direct result of or are related to organisational structure, corporate culture, management style and other firm-specific characteristics of MNEs. The emphasis on these characteristics appears to have been influenced by the particulars of the studies' samples. Most of the previous studies were based on either a sample of MNEs from a specific country, mainly the United States, or on a small sample of firms from different industries. These sampling limitations have restricted the scope and applicability of the findings, and could not possibly capture and demonstrate the diversities or similarities in centralisation for various MNEs of different industries and countries.

This study is designed to examine the centralisation of technological development in MNEs and its associated similarities or diversities at the MNEs' industry and home country levels. The authors, however, do not refute the importance of firm-specific characteristics in influencing or determining the location or degree of R&D centralisation at the firm level.[2] Instead, this study's findings will supplement these characteristics in a limited and narrow way.

THE USE OF PATENTS AS AN INDICATOR OF MNEs' INVENTIVE ACTIVITY

To study the MNEs' pattern of similarities and differences in the development of their own technological environment, this study examines their patenting activity.[3] Patent-related information was extracted from the PATDAT data bank of Consumer and Corporate Affairs Canada.[4] These data include the inventors' country of residence, the patentees' name and country and the corporate links between firms on a worldwide basis. The data contain exact and complete information on the patenting activity of foreign subsidiaries, home country subsidiaries and the MNEs' corporate headquarters. The data also provide useful insights into their R&D activity. Assuming that all patents granted to an MNE are the results of the MNE's inventive activity and that, for each patent, the inventor's country of residence identifies a subsidiary in that country as the origin of the invention, the data supply information on the foreign and domestic R&D activity of MNEs.

SAMPLE AND CHARACTERISTICS OF MNEs

The first objective in selecting the sample of MNEs was to have a good representation of North American, European and Japanese-based MNEs. The second objective was to select MNEs from industries with different characteristics. For these reasons, Stopford and Dunning's Directory of Multinationals (1983) was used to generate a list of MNEs whose headquarters are based in the USA, Canada, Europe and Japan with main activities in four broad industrial sectors.

In selecting the sample, close attention was paid to a variety of concerns based on the findings of previous research or observed industrial behaviour. In explaining the differences in the centralisation of various industries' R&D activity (at the home country or corporate headquarters), previous studies point to the importance of technology as being the dominant factor. A majority of the studies postulated a positive link between the industry's technological intensity and the degree of centralisation of technological development in the MNE's home country.[5] However, other industry-specific characteristics were found to explain inter-industry differences in the centralisation of R&D. A greater necessity in some industries to adapt their products to local market conditions or host government regulations has led to a higher degree of decentralisation in those industries.[6] The needs of process development, as opposed to product development, also favour decentralisation.[7]

In making the choice of these four sectors, one final consideration was also satisfied. Namely, industries in which patents are known to be an imprecise indicator of R&D activity were excluded. For example, automobile and defence-related industries are industries in which patents are not a good indicator of R&D activity (Pavitt, 1982).

These concerns and requirements led to the selection of four sectors: industrial and agricultural chemicals, electrical engineering and electronics, industrial and farm equipment, and metals and metal products. The first three sectors are reported to be highly active in R&D abroad (Industry Studies Group, 1979).

Stopford and Dunning's Directory listed 203 MNEs in the four sectors for which consolidated sales and employment figures of 188 and 176 MNEs were respectively listed.[8] PATDAT data, however, identified 197 MNEs with patent holdings during the 1980–83 period. As foreign sales and foreign employment of 45 MNEs were not listed in the Stopford and Dunning Directory, sample size was further reduced.[9] Finally, for the joint consideration of patents with foreign employment, the sample size was reduced to 133.[10]

Selected characteristics of the MNEs included in the sample are presented in Table 7.1. The average sales figures of the 188 MNEs is $US 2.64 billion. A comparison of the share of foreign employment among the North American and European MNEs shows that the latter have a much stronger presence in foreign markets. With the exception of the chemical sector, North American MNEs are usually the most technologically active since they show the greatest technological intensity. On average, European MNEs hold fewer patents relative to their size than North American or Japanese MNEs.

Foreign subsidiaries' degree of participation in the MNEs' total inventive activity varies across sectors and countries. The share of foreign patented inventions is at its highest in the equipment sector (22 per cent) while it is at its lowest in the electrical sector (10 per

Table 7.1: MNEs' inventive activity and characteristics

	Chemicals	Electrical	Equipment	Metals	All sectors
Sample of 197 MNEs:					
Total number of MNEs	48	41	45	63	197
North American MNEs	26	18	20	21	85
European MNEs	17	12	19	27	75
Japanese MNEs	5	10	5	11	31
Average number of patents granted to:					
All MNEs	63.2	56.1	19.3	11.7	35.2
North American MNEs	54.8	61.0	30.6	12.6	38.9
European MNEs	86.7	53.0	10.1	12.1	35.1
Japanese MNEs	26.8	50.4	7.2	8.5	24.7
Average share of patents for foreign inventions:					
All MNEs	0.14	0.10	0.22	0.12	0.14
North American MNEs	0.07	0.09	0.15	0.12	0.11
European MNEs	0.28	0.17	0.34	0.17	0.24
Japanese MNEs	0.00	0.02	0.05	0.00	0.01
Sample of 188 MNEs:					
Average worldwide sales (in billion US dollars)					
All MNEs	3.27	3.54	1.55	2.26	2.64
North American MNEs	2.88	3.15	1.88	2.00	2.46
European MNEs	4.45	4.01	1.22	2.39	2.86
Japanese MNEs	1.41	3.62	1.48	3.28	2.84
Average number of patents granted to MNEs/average worldwide sales					
All MNEs	19.9	15.5	12.6	5.3	13.8
North American MNEs	19.3	19.8	16.3	6.0	15.8
European MNEs	19.5	11.3	7.9	4.9	11.8
Japanese MNEs	22.3	15.4	5.0	3.8	12.8

Sample of 133 MNEs:					
Worldwide employment (in thousands)					
All MNEs	57.6	107.6	34.4	42.2	58.4
North American MNEs	42.3	83.8	36.4	34.2	48.4
European MNEs	83.1	144.2	31.8	50.3	72.3
Average foreign employment/average worldwide employment					
All MNEs	0.32	0.30	0.31	0.30	0.31
North American MNEs	0.23	0.25	0.29	0.25	0.25
European MNEs	0.47	0.37	0.35	0.34	0.38

Source: Tables and charts in this paper were derived from data for 1980–83 collected by the authors, as described in the text.

cent). On average, foreign subsidiaries' share of patented inventions stands at 14 per cent. For American, European and Japanese-based MNEs, there exist differences in the share of foreign-patented inventions originating at the subsidiary level. For the European MNEs, foreign inventions average 24 per cent of total patent holdings. This indicates a much greater degree of decentralisation in foreign countries for European MNEs. The foreign inventions of North American and Japanese MNEs equal 11 per cent and 1 per cent respectively. For Japanese MNEs, almost all of their inventive activity is centralised in the home country for every sector. At the sectoral level, differences in the foreign subsidiaries' participation in the total inventive activity of North American and European MNEs are less important in the metal sector than in the equipment, chemical and electrical sectors.

THE DECENTRALISATION OF TECHNOLOGY DEVELOPMENT IN MNEs

MNEs can conduct their inventive activity in various subsidiaries or concentrate it in the home country or even at headquarters level. The objective of the analysis is to examine whether the differences in the degree of decentralisation of technological development are related to the MNEs' sector of activity and/or their country of origin.

Numerous studies have estimated the MNEs' share of foreign R&D activity.[11] In contrast with these studies, the present one uses information on the inventor's country of residence to estimate the shares of home and foreign countries' patented inventions. However, as mentioned previously, it is necessary to assume that a foreign country's patented inventions and hence its share of the total patents granted to an MNE are the direct result of, or are related to, inventions of the MNE's subsidiary in that country. Therefore, in the light of this assumption, the number of patents for foreign country inventions can be used to measure the extent of decentralisation of the R&D activity of MNEs.

Empirical studies have shown that the foreign subsidiaries' share of the MNEs' total inventive activity varies among MNEs of different countries of origin. When compared to American MNEs, the inventive activity of European MNEs is more decentralised, while Japanese inventive activity is very centralised. These facts have been well documented by numerous studies.[12] The greater decentralisation among European MNEs is explained by the proximity of large foreign European markets, the attraction of the large American market and the relatively small size of their home market. On the contrary, the greater — if not complete — centralisation of R&D activity of Japanese MNEs is linked to their recent internationalisation, their low foreign production involvement and low domestic R&D costs.

Theoretical studies have broadened the list of determinants for centralisation and decentralisation of R&D in foreign subsidiaries.[13] Among the factors favouring centralisation is the critical mass. A minimal critical mass at R&D centres is required to allow for economies of scale, for the benefits of complementarities and externalities of inventive activity within a centre and to facilitate badly-needed communication and co-ordination among the centres. One of the factors that has an impact on the decentralisation decision of R&D centres is the need to adapt products and processes to local markets and host country regulations.

The minimal critical mass guarantees a certain efficiency level of the inventive activity. Any R&D centre's level of activity must reach or exceed the minimal critical mass before it can be decentralised. However, the advantages and benefits of responding to foreign markets' needs and requirements cannot begin to influence the decentralisation decision before foreign production levels reach a certain degree of importance (e.g. a minimum share of MNE's overall production). Therefore, one would expect an interaction between the minimal critical mass and the share of foreign production.

The decision to decentralise R&D activity in an MNE is, in fact, composed of two decisions: first, whether to do research abroad, and, second, the amount of funds that should be allocated to foreign research. The TOBIT model is well suited to handle this situation. This model is based on the assumption that the dependent variable has a number of its values clustered around a limited value — zero in the case of patents for foreign inventions. The incidence in the sample of no foreign inventions, and hence no patented foreign inventions, is relatively frequent. In fact, 50 MNEs had not taken out any Canadian patents for foreign inventions during the 1980–83 period, while the remaining 147 had at least one patent for inventions of foreign origin.

To capture the impact of the minimal critical mass requirement, the TOBIT model is used to regress the number of patented foreign inventions, P_f, as a function of the total number of patents during the 1980–83 period, P (see Table 7.2). For comparison, the results

Table 7.2: Share of patented foreign inventions in the MNEs' overall patenting activity

Specifications	Estimated coefficients a_0	a_1	a_2	a_3	a_4	R^2	F	L	N
$P_f = a_0 + a_1P$									
OLS	−1.58	0.160				0.52	162.1*	−744.9	147
	(−0.41)	(12.73)*							
		[1.06]							
TOBIT	−13.92	0.177						−779.7	197
	(−3.95)*	(11.37)*							
		[0.70]							
$P_f = a_0 + a_1P + a_2DEUP + a_3DJPP$									
OLS	1.98	0.072	0.180	−0.060		0.79	183.0*	−684.1	147
	(0.76)	(6.28)*	(12.59)*	(−1.97)**					
		[0.47]	[0.48]	[−0.03]					
TOBIT	−6.37	0.087	0.177	−0.083				−718.7	197
	(−2.73)*	(7.21)*	(9.93)*	(−2.85)*					
		[0.46]	[0.35]	[−0.05]					
$P_f = a_0 + a_1P + a_2DELP + a_3DEQP + a_4DMEP$									
OLS	−2.14	0.181	−0.040	−0.074	0.064	0.53	42.7*	−742.0	147
	(0.49)	(11.15)*	(−1.87)**	(−1.40)	(0.68)				
		[1.20]	(−0.09)	[−0.07]	[0.04]				
TOBIT	−15.42	0.201	−0.049	0.026	0.060			−776.7	197
	(−3.88)*	(10.46)*	(−2.27)**	(−0.50)	(0.66)				
		[0.81]	[−0.06]	[−0.01]	[0.03]				

Notes: The t statistics of the regression coefficients are given in parentheses. The elasticities at the mean value of the variables are given in brackets. The single asterisk (*) and the double asterisk (**) indicate that the estimated coefficients or statistics are respectively significant at the 0.01 and 0.05 levels.
[...] log likelihood function and N, number of observations.

of the ordinary least square (OLS) regressions are also reported. It is important to note that the cases with no patented foreign inventions are dropped from the OLS regressions. Furthermore, the various elasticities of patented foreign inventions for both models are given in the tables. The overall results provided by the TOBIT model indicate that the model works quite satisfactorily. It shows that, as the total number of patents granted to an MNE approaches the infinite value, the share of patented foreign inventions asymptotically reaches 18 per cent.

The TOBIT model decomposes the MNEs' expected number of patented foreign inventions into two components: the probability that MNEs will have patented foreign inventions and the expected number of patented foreign inventions for MNEs already involved in foreign inventive activity. Both the probability of conducting foreign inventive activity and the expected number of patented foreign inventions increase with the total number of patents granted to MNEs.

At the mean value of P (141 patents for an MNE over the four-year period), the elasticity of the probability of having patented foreign inventions equals 0.39, while the elasticity of the expected number of patented foreign inventions for MNEs above the limit equals only 0.28. This means that, at the mean value of P, the changes in the total number of patents granted to MNEs are still principally translated into increases in the total number of MNEs that are beginning to have patented foreign inventions. It is only in the vicinity of 200 patents that the relative importance of the two effects begins to reverse: then, the increases in the total number of patents granted to MNEs are predominantly associated with increases in the expected number of patented foreign inventions for MNEs already involved in foreign inventive activity.

The introduction of dummy variables for the regions of origin (i.e. DEU and DJP, respectively, for European and Japanese-based MNEs) into the TOBIT regression points to highly significant differences between the North American, European and Japanese MNEs.[14] Conversely, the introduction of sectoral dummy variables (i.e. DEL for electricals, DEQ for equipments and DME for metals) fails to improve the global fit of the estimation. However, they show a slight difference in the decentralisation share of electricals as compared to chemicals (15 per cent as opposed to 20 per cent).

The share of patented foreign inventions shows a tendency towards regional differences: 26 per cent for the European MNEs and 9 per cent for the North American MNEs. The Japanese MNEs

Figure 7.1 A TOBIT analysis of patented foreign inventions and the MNEs' overall patenting activity

$$P_f = a_o + a_1P + a_2DEUP + a_3DJPP$$

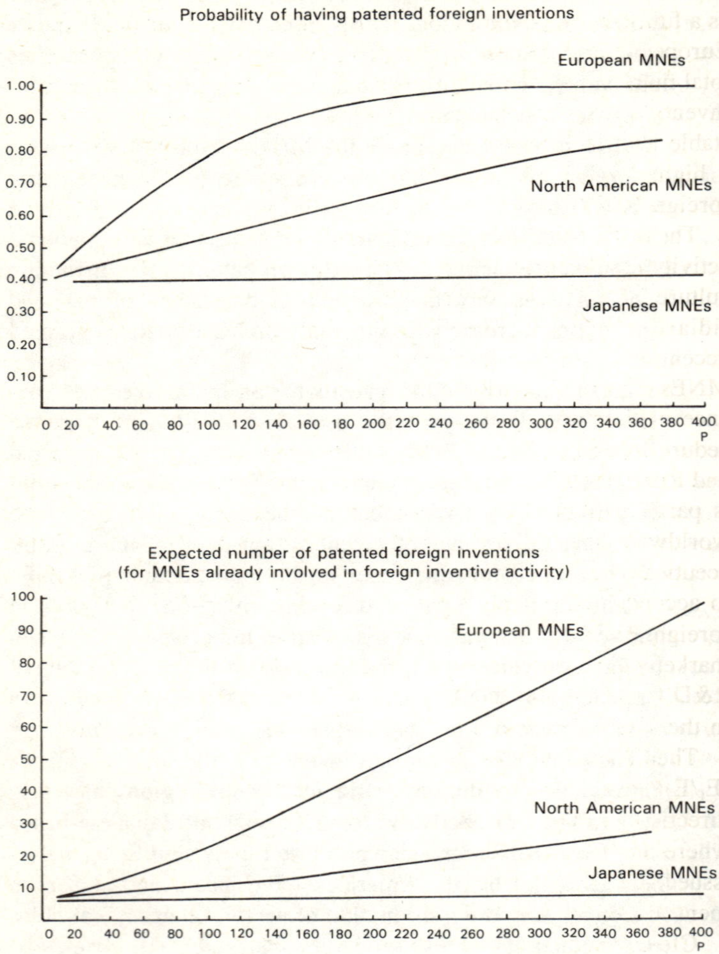

do not take part in foreign inventions patenting activity. Regional differences are statistically significant in general, and even more so for the European MNEs.

Figure 7.1 depicts the differences in the probability of having patented foreign inventions and the expected number of such patents as a function of the total number of patents for the North American, European and Japanese MNEs. Figure 7.1 also shows that, as the total number of patents increases, the probability of having foreign inventive activity in subsidiaries is initially low and remains low and stable for the Japanese MNEs. In contrast, European MNEs show a higher initial and rapidly rising probability of being active in foreign R&D.

These differences in the decentralisation of the MNEs' inventive activity can be attributed to regional dissimilarities in the managerial culture, the MNEs' organisational structure and the age of subsidiaries.[15] The dissimilarities in the patterns of R&D decentralisation among the North American, European and Japanese MNEs can also be attributed to the differences in the relative importance of their foreign production. A study based on the OLS procedure applied to a sample of 55 American MNEs (Mansfield, Teece and Romeo, 1979) shows that the share of R&D funds spent abroad is partly explained by the percentage of their foreign sales, their worldwide sales, and a dummy variable for MNEs in the pharmaceutical industry. These three explanatory variables were included to account respectively for the importance of foreign markets and foreign R&D activity geared to the special design needs of foreign markets; the economies of scale and minimal efficient size of foreign R&D facilities; and the foreign R&D activities linked to regulation in the pharmaceutical industry.

The TOBIT regression of P_f as a function of E_f/E, P and $(E_f/E)P$, estimated by the following specification, points to a clear direction (Table 7.3): $P_f = a_0 + a_1 (E_f/E)P + a_2(E_f/E) + a_3P$, where P_f = patented foreign inventions and P = total patents issued to MNEs during the 1980–83 period; E_f = foreign employment and E = worldwide employment of MNEs in 1977.

Clearly not only is $(E_f/E)P$ the most highly significant coefficient, but P and E_f/E fail to produce coefficients with any statistical significance. A chi-square likelihood ratio test confirms that E_f/E and P do not add to the explanatory power of $(E_f/E)P$. This leads to the adoption of an interactive form as the sole specification.

To test the joint influence of the minimal critical mass require-

Table 7.3: Comparison between MNEs' share of foreign inventive activity and MNEs' share of foreign employment

Specifications	Estimated coefficients								
	a_3	a_1	a_2	a_3	a_4	R^2	F	L	N
$P_f = a_o + a_1(E_f/E)P$									
OLS	2.32 (0.85)	0.421 (22.89)* [0.93]				0.82	524.1*	−552.1	118
TOBIT	−1.16 (−0.44)	0.430 (13.01)* [0.79]						−564.2	133
$P_f = a_0 + a_1(E_f/E)P + a_2DEU(E_f/E)P$									
OLS	7.02 (2.83)*	0.226 (6.51)* [0.50]	0.228 (6.27)* [0.28]			0.86	368.3*	−534.8	118
TOBIT	3.31 (1.37)	0.247 (6.52)* [0.48]	0.214 (5.44)* [0.23]					−549.3	133
$P_f = a_0 + a_1(E_f/E)P + a_2DEL(E_f/E)P + a_3DEQ(E_f/E)P + a_4DME(E_f/E)P$									
OLS	1.06 (0.34)	0.430 (17.24)* [0.95]	−0.017 (−0.50) [−0.01]	−0.041 (−0.32) [−0.01]	0.244 (1.51) [0.04]	0.82	131.6*	−550.6	118
TOBIT	−3.08 (−1.06)	0.443 (11.88)* [0.82]	−0.023 (−0.07) [−0.01]	0.019 (0.15) [0.00]	0.294 (1.84)** [0.04]			−562.2	133

$P_f = a_0 + a_1(E_f/E)P + a_2(E_f/E) + a_3P$

OLS	2.11	0.48	0.08	0.03		0.82	180.9*	−549.6	118
	(0.39)	(10.98)*	(0.61)	(1.69)**					
		[1.05]	[0.09]	[0.20]					
TOBIT	−5.68	0.45	0.21	−0.02				−561.2	133
	(−1.15)	(8.63)*	(1.63)	(−1.04)					
		[0.85]	[0.20]	[−0.10]					

Notes: For symbols, see notes for Table 7.2.

The delay between the completion of the invention and patent issuing is about five years. Therefore the MNEs' patent holdings during the 1980–83 period (average date 01/01/82) are studied in relation to the 1977 employment figures.

ment and the pressure to adapt products and processes to the needs of foreign markets, the TOBIT regression procedure is used to estimate the following specifications: $P_f = a_0 + a_1 (E_f/E)P$.

The term $(E_f/E)P$ gives an estimate of the MNEs' share of patented foreign inventions, should foreign inventive activity be proportional to the share of foreign employment. A relatively low level of total inventive activity and a relatively low share of foreign employment is not conducive to inventive R&D activity in foreign subsidiaries. Conversely, a joint increase in $(E_f/E)P$ may result in further decentralisation and a higher number of foreign inventions. The TOBIT estimation shows that the observed number of patented foreign inventions approaches only 43 per cent of the hypothetical number which is proportional to the share of employment in foreign subsidiaries. These results point to a greater centralisation of R&D than of production in MNEs.

The cumulative probability of having patented foreign inventions and the expected number of such patents increases rapidly with the hypothetical number of foreign inventions. At the mean value of the hypothetical number of patented foreign inventions (which is equal to 61 over the four-year period), the elasticity of the expected number of patented foreign inventions outweighs the elasticity of the probability of having patented foreign inventions. This result indicates that, at the mean, an increase in the total number of patented foreign inventions comes predominantly from further activity of the MNEs already active in foreign inventive activity. The introduction of dummy variables for regions of origin in order to compare and contrast the North American and European MNEs reveals some interesting results.[16] Results in Table 7.3 and Figure 7.2 show that when $(E_f/E)P$ begins to increase, the European MNEs respond more rapidly by further decentralising their inventive activity in the direction of their foreign subsidiaries than do the North American MNEs.

The introduction of dummy variables for the regions of origin in the TOBIT regression of $P_f = a_0 + a_1P$ produces significant coefficients and adds to the goodness-of-fit of the original regression without dummies (log likelihood function changed from −779.7 to −718.7). Although the same treatment in the TOBIT regression of $P_f = a_0 + a_1(E_f/E)P$ matches these results and the dummy variables' coefficients for regions of origin are still significant, the regional differences for the latter model are much less pronounced (for a comparison, see Tables 7.2 and 7.3 and Figures 7.1 and 7.2). This suggests that in the interactive term, E_f/E embodies part of the

Figure 7.2 A TOBIT analysis of MNEs' share of foreign inventive activity and MNEs' share of foreign employment

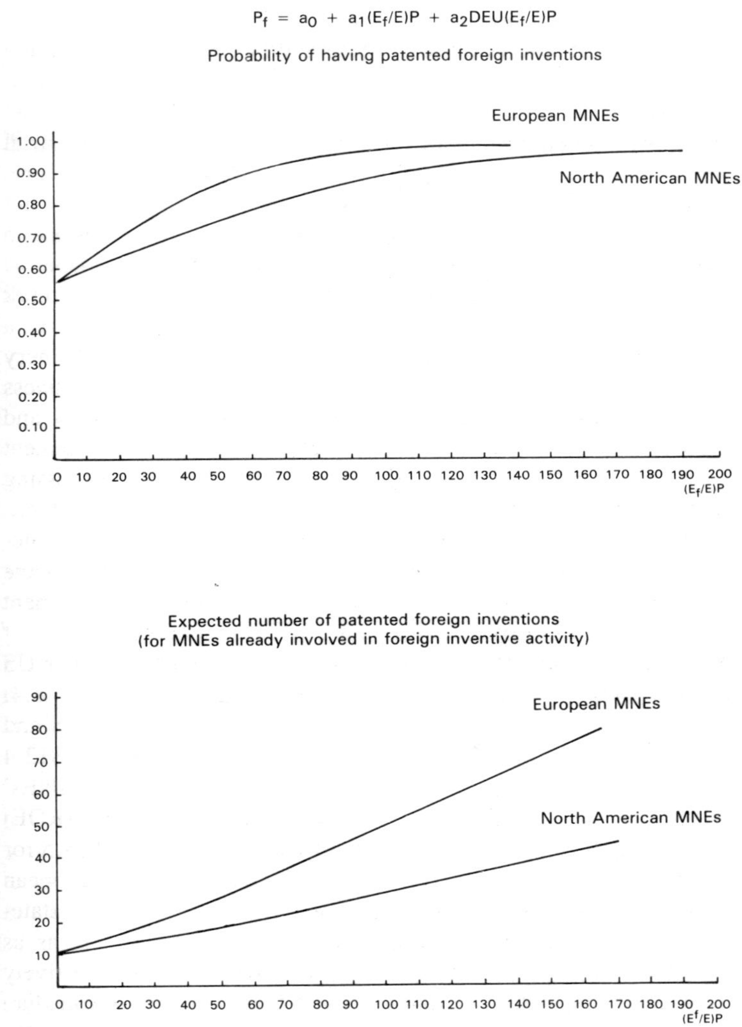

explanatory power of the regional variations and hence reduces the importance of the dummy variables for the regions of origin. Nevertheless, besides the greater importance of foreign production for European MNEs, other factors also seem to be responsible for the greater decentralisation of inventive activity.

Although the introduction of sectoral dummy variables does not improve the overall statistical results, it conclusively shows a larger decentralisation for metals in comparison to chemicals. It also proves that metals asymptotically reach a greater decentralisation than chemicals at higher levels of $(E_f/E)P$. Conversely, metals do not show much decentralisation at lower levels of $(E_f/E)P$, as there are nine firms (out of the 34 firms in the metal sector) showing no patented foreign inventions ($P_f = 0$). These results indicate a dichotomy within the metal sector: there is no decentralisation at lower levels while there is larger decentralisation at higher levels. A plausible explanation could be the lower technological intensity and the greater process orientation of metals as compared to chemicals.[17]

CONCLUSION

Several ambiguous areas in the literature provided the initial motivation for this study. The primary objective was to explore the pattern of variations in the decentralisation of inventive activity in terms of MNEs' overall inventive activity, extent of foreign production, main sector of activity and region of origin. Although the literature on inventive activity and patents covers various facets of the subject, the simultaneous consideration of an MNE's size, main sector of activity and region of origin has not been addressed.

The literature covering the organisational aspects of inventive activity suggests a conceptually appealing postulate: a higher degree of technological intensity is coupled with a greater extent of centralisation of inventive activity across firms. The diversity and richness of this study's data set, including the wide range of variations in the technological intensity of MNEs, were highly conducive to an examination of the various facets of the relationship between technological intensity and decentralisation. This study's findings are not categorically supportive of the postulate above; the main sectors of activity fail to show much difference when compared with chemicals, for which technological intensity is the highest of the four sectors. The failure to confirm technological intensity as the sole explanatory factor in decentralisation of inventive activity led the authors to a further explanation of a selected number of factors and specification forms.

The level of overall inventive activity appears to play a special role in decentralisation: first, at the low levels, the probability of any decentralisation is very weak; and second, as the levels reach higher values, the projected foreign inventive activity reaches 18 per cent asymptotically. Additionally, MNEs' regions of origin account for significant differences in the patterns of decentralisation. The European MNEs show the greatest tendency to decentralise their inventive activity. While the North American MNEs follow the European, Japanese MNEs show the least tendency to decentralise.

The combined effect of the inventive activity's overall size and pressures to decentralise in order to adapt to local needs along with shares of production and employment in foreign subsidiaries appears to explain the level of decentralisation. This combined effect seems to embody some of the regional variations to the extent that regions of origin lose some of their explanatory power in the interactive specification. Again, the main sectors of activity failed to improve the results or achieve reasonable statistical significance. Inventive activity is less decentralised than production and employment; compared with the share of the other two, only 43 per cent of inventive activity is conducted in foreign subsidiaries.

The emerging implication of this study is that factors influencing the conduct of an MNE's inventive activity and its associated decision — i.e. decentralisation — are integral parts of an overall competitive strategy. Although this study examined the effects of size, region of origin and main sector of activity on the pattern of overall inventive activity and its decentralisation, a host of other influences form a long agenda for future studies.

NOTES

1. Notes of thanks and appreciation are due to Claude Desranleau, Yves Fortier, Hani Zayat and Kathleen Deslauriers for their assistance in the different phases of this study.

2. For a detailed study of the influence of firm-specific characteristics on the centralisation of R&D and patenting activities, see Etemad and Séguin Dulude (1984, 1985).

3. The strongest argument against the use of patents as an indicator of inventive activity is the existence of 'trade secrets'. Nevertheless, MNEs are known to attach a great deal of importance to their patenting activity. In an

extensive study on the patenting activity of 82 multinational corporations (MNCs), Wyatt (1984, p. 100) concluded that: 'Since the propensity to patent in capitalist, industrialized countries is quite high, comparison of numbers of patents obtained in such countries provides some indication of a MNCs relative (to other MNCs) technological performance, particularly within a sector'.

4. The authors would like to acknowledge and thank *Consumer and Corporate Affairs Canada* for its invaluable help in allowing them access to information on patents granted by the Government of Canada. The analysis, conclusions and views expressed in this study, however, are solely the authors' responsibility.

5. See on this point, Behrman and Wallender (1976); de Bodinat (1984); Fischer and Behrman (1979); Hakansson and Laage-Hellman (1984); Hewitt (1980); and Stobaugh and Telesio (1983).

6. Studies by Behrman and Fischer (1980a,b); Hakansson and Laage-Hellman (1984); Robock and Simmonds (1983); Terpstra (1977, 1983) emphasise this factor. Pharmaceuticals and chemicals are sectors often mentioned.

7. A factor mentioned by Coughlin (1983).

8. In the Stopford and Dunning Directory list, there is a total of 205 MNEs. Two MNEs were then excluded since one MNE is classified and listed in two selected industries, and another has corporate links incompatible with the information contained in PATDAT.

9. In the Stopford and Dunning Directory, data on shares of both foreign sales (S_f) and foreign employment (E_f) were available for 64 MNEs. From this sample, a regression was estimated: $S_f = 0.23 + 1.00 E_f$ ($R^2 = 0.70$, $F = 147.2$. The t statistics for the two coefficients were respectively 0.08 and 12.13). For the remaining MNEs, with either data on S_f and E_f, the missing share was estimated and substituted on the basis of the equation.

10. The number of Japanese-based MNEs in the 133 sample being too small, they were excluded from this sample.

11. See on this point, Behrman and Fischer (1980a); Creamer (1976); Hakansson and Laage-Hellman (1984) reporting Swedenborg's study; Industry Studies Group (1979); Lake (1979); Mansfield, Teece and Romeo (1979); and Ronstadt (1977).

12. See on this point, Behrman and Fischer (1980a,b); Business International (1971); Duerr (1970); Franko (1976); Ronstadt (1977); Ronstadt and Kramer (1982); and Terpstra (1977, 1983).

13. See Etemad and Séguin Dulude (1984); Mansfield, Teece and Romeo (1979); and Terpstra (1977).

14. A chi-square likelihood ratio test which was significant at the 0.01 level clearly confirms the positive contribution of the dummy variables for the regions of origin.

15. The explanatory power of the last two factors was in fact successfully tested with a sample of American MNEs by Hewitt (1980) who had access to the 'Harvard University Multinational Enterprise Project' data base.

16. A chi-square likelihood ratio test significant at the 0.01 level indicates that regions of origin contribute significantly to the explanatory

power of the specified model.

17. This area presents itself as a promising subject for future research with respect to the metal sector.

8

Canadian Foreign Direct Investment

Alan M. Rugman

INTRODUCTION

From a Canadian perspective over the past ten years traditional patterns of foreign direct investment (FDI) have been reversed. Inflows of FDI into Canada (mainly from the United States) have been replaced by surprisingly large outflows of Canadian FDI, again predominantly into the United States. This paper presents recent data on the flows of FDI, examines the reasons for the dramatic reversal, draws out theoretical lessons from an analysis of the major foreign investors and, finally, offers some suggestions for the strategic management of Canada's large multinationals.

The focus of the paper is not only upon the aggregative data on inward and outward flows of FDI. Rather, the latter portion of the paper moves to a more micro-level analysis, case by case, of the actual multinational enterprises (MNEs) undertaking the outward FDI. A group of the 16 largest Canadian MNEs is identified and these are studied in detail. It is found that most of these MNEs do not have the traditional high-tech firm-specific advantages of large US, European and Japanese MNEs. Instead, they are frequently in mature, resource-based industries where their advantages are stronger on the marketing end of the business, rather than on the production end. The reasons for the strategic success of these MNEs will be drawn out and generalised so that the theory of the MNE and the analysis of technology transfer can be somewhat modified to take account of the Canadian phenomenon.

CANADA AS AN OUTWARD INVESTOR

It has become apparent to citizens of Canada and the world at large that we live in an interdependent world economy. Today we witness massive movements of goods and financial flows between nations. Nowhere is this international exchange by trade and investments more interlinked than between Canada and the United States.

For Canadians, the experience of FDI has been conditioned by the nature of foreign ownership of the economy. For the last half century some 80 per cent of all inward FDI has been from the United States and US ownership of the Canadian economy has become a sensitive political issue. Previous studies have examined the nature, economic causes and policy aspects of United States FDI in Canada, see Globerman (1979) and (1985) and Safarian (1969). Another study, Dunn (1978), placed the special role of FDI into a perspective against an analysis of the balance of payments accounts of the United States and Canada. His work is especially strong in applying financial analysis to an understanding of the capital account. Shapiro (1980) has examined the performance of foreign-owned firms in Canada. A related piece with a similar financial focus is Pattison (1978).

Canada has traditionally relied upon net inflows of FDI to help finance its real economic development. On occasions there have also been net inflows of long-term portfolio capital (over which investors do not exert control as they do with FDI). Short-term capital flows have traditionally been very volatile, responding to interest rate differentials between Canada and the United States. These relationships are discussed in detail in Rugman (1980a), especially in Chapter 10. In turn, this work is an extension of the classic study of foreign ownership in Canada by Safarian (1966).

The focus of this paper is Canada's outward FDI. Starting in 1975 Canada has reversed its traditional pattern of dependence on US inward FDI and turned instead to become a net outward investor, mainly in the United States. This will now be examined and some preliminary analysis of these figures will be undertaken. A detailed explanation of Canada's changed nature in the realm of foreign direct investment appears in Rugman (1986), especially in Chapters 2 and 3 on the economic and managerial determinants of FDI respectively. Not only Canadian FDI has been attracted to the United States; many European and other multinationals are going there. This phenomenon of the United States turning from the world's largest outward source of FDI to a net importer of FDI in recent years is examined in Gray (1985). Earlier studies on inward FDI into the United States were by Arpan and Ricks (1974) and Webley (1974).

Canada: from capital importer to capital exporter

The historical nature of Canada's FDI flows is reported in Table 8.1. The three columns of this table are worthy of study.

Capital inflows

Over the last 25 years there has usually been an annual inflow of FDI into Canada, as shown in column (1). This averaged over half a billion dollars per year until 1976 when we witnessed the first instance of repatriation of FDI. Following this there were four more years of inflows of FDI until a sudden and major repatriation of FDI occurred in 1981. This continued into 1982, but the last two years once again brought inflows of FDI. This column confirms Canada's historical reliance on inward foreign direct investment, leading to substantial foreign ownership of the Canadian economy. It also indicates that in recent years this pattern of capital dependency has been disturbed on several occasions.

Most of this inward FDI has been from the United States (about 80 per cent on average), leading to political concerns over the extent and concentration of US ownership of the Canadian economy. These concerns have been expressed in numerous federal government studies such as the Watkins Report (1968), the Wahn Report (1970), and the Gray Report (1972).

Capital outflows

The persistent outflow of FDI from Canada is reported in column (2) of Table 8.1. In each year of the last quarter century there have been such outflows. Over the first half of this period the outflows of Canadian FDI were small, averaging about $170 million per year. In the early 1970s this increased to about $400 million per year, and since 1978 there have been major outflows averaging over three billion dollars annually, except for 1982 when it was about one billion.

Most of these Canadian outflows of FDI have gone to the United States (about 70 per cent of the total in 1980). However, this has not led to US concerns about Canadian ownership of its economy, since the Canadian inflows have a relatively small (indeed trivial) impact

Table 8.1: Canadian foreign direct investment flows, 1960–84 (C$ millions)

	(1) Foreign direct investment in Canada	(2) Canadian foreign direct investment abroad	(3) Net flows [(1) + (2)]
1960	+ 670	− 50	620
1961	+ 560	− 80	480
1962	+ 505	− 105	400
1963	+ 280	− 135	145
1964	+ 270	− 95	175
1965	+ 535	− 125	410
1966	+ 790	− 5	785
1967	+ 691	− 125	566
1968	+ 590	− 225	365
1969	+ 720	− 370	350
1970	+ 905	− 315	590
1971	+ 925	− 230	695
1972	+ 620	− 400	220
1973	+ 830	− 770	60
1974	+ 845	− 810	35
1975	+ 725	− 915	− 190
1976	− 300	− 590	− 890
1977	+ 475	− 740	− 265
1978	+ 135	−2325	− 2190
1979	+ 750	−2550	− 1800
1980	+ 800	−3150	− 2350
1981	−4400	−6900	−11300
1982	− 900	− 950	− 1850
1983	+ 200	−2700	− 2500
1984	+2380	−4025	− 1645

Notes: Data exclude reinvested earnings.
A minus sign indicates an outflow of capital from Canada; it represents an increase in holdings of assets abroad or a reduction in liabilities to non-residents.
Source: Statistics Canada, *Quarterly Estimates of the Canadian Balance of International Payments:* Catalogue No. 67–001, Table 40.

on the United States. This is the first example of the US-Canadian size asymmetry which lies at the heart of analysis of bilateral economic relations. If there were approximately equal capital flows between the nations clearly a one billion dollar US inflow into Canada would have a much greater impact than an identical Canadian inflow into the United States.

Net capital flows

The dramatic reversal of net capital flows in 1975 is illustrated in column (3) of Table 8.1. Before then the net inward flows averaged nearly $400 million a year. Since then net outflows have averaged six times this at $2.5 billion a year. Canada's new status as a net capital exporter over the last ten years represents a dramatic change from its role as a net capital importer for the previous 15 years, and longer.

In Rugman (1986) the reasons for this reversal in Canada's FDI position are examined. It is found that 1975 represented a critical structural break for Canada but that the massive net capital outflow of over eleven billion dollars in 1981 was part of a general upsurge of FDI into the United States by the advanced industrialised nations. This upsurge is a type of intra-industry FDI in which MNEs engage in cross-investments in each other's home markets.

To a large extent Canadian FDI in the United States over the 1979–84 period has been tracking that of other nations. The US market appears to be the major variable attracting such inward direct investment; MNEs want to overcome US barriers to trade and retain access to the world's single largest market. In Canada's case such pull factors appear to be more important than specific push factors, except for the general fact that Canadian MNEs are now supported by a more mature domestic economic and financial system, which helps them to make FDI decisions as part of their overall strategic planning. Studies of the motives for FDI at the firm level, reported in Rugman (1986), also indicate that market accessibility and other pull factors are more important determinants of Canadian outward FDI than are push factors.

The reversal of Canada's investment stock position

The reversal in Canadian direct investments is also confirmed by analysis of stock figures, as in Table 8.2 on Canada's investment position. These Statistics Canada figures on the book value of Canada's direct investment position in the United States (column (1)) reveal a doubling of the stock in the last five years, up to nearly $30 billion by 1984. While the US direct investment position in Canada has also increased in this period (shown in column (2)) the value of this stock has risen only at a much smaller rate.

The Canadian position in the United States as a percentage of the US position in Canada has shown a remarkable change, from some 19 per cent in 1975 to 28 per cent in 1979 to over 46 per cent by 1984. Even if allowance was made for the greater age of the US stock, and especially its undervaluation in these book value terms,

Table 8.2: Bilateral stocks of foreign direct investment by Canada and the United States, 1975–84 (C$ millions)

	Canadian FDI position in the US	United States FDI position in Canada	Net position	Canada/US %	US/Canada %
1975	5,559	29,666	−24,107	18.7	553.6
1976	6,092	31,917	−25,825	19.1	523.9
1977	7,116	34,720	−27,604	20.5	487.9
1978	8,965	38,348	−29,383	23.4	427.8
1979	12,104	42,771	−30,667	28.3	353.4
1980	16,378	48,684	−32,306	33.6	297.3
1981	21,832	52,300	−30,468	41.7	293.6
1982	22,990	53,600	−30,610	42.9	233.1
1983	25,027	57,400	−32,373	43.6	229.4
1984	29,629	64,210	−34,581	46.1	216.7
Average rate of increase 1975–84	19.7%	8.9%			

Source: Statistics Canada, *Canada's International Investment Position 1979 and 1980* (Ottawa, 1984, Catalogue No. 67–202, Table 2 and Table 18). Data for 1981–4 supplied by the International Investment Position of the International and Financial Economics Division of Statistics Canada.

Table 8.3: Value and shares of foreign direct investment in the United States, 1984 and 1975

	Value at year end (US $ millions)	1984 per cent	1975 per cent	Per cent change
Canada	14,001	8.8	19.3	(54.4)
Europe	106,567	66.8	67.2	(0.6)
France	6,502	4.1	4.9	(16.3)
Germany	11,956	7.5	5.1	47.1
United Kingdom	38,099	23.9	22.9	4.4
Netherlands	32,643	20.5	19.3	6.2
Japan	14,817	9.3	2.1	342.9
Other Western hemisphere	12,711	8.0	8.0	0.0
Netherlands Antilles	10,523	6.6	na	na
Rest of world	13,840	8.7	2.0	335.0
Middle East	5,159	3.2	0.8	300.0
Australia, New Zealand South Africa	2,366	1.5	0.1	1,400.0
Other Africa, Asia and Pacific	1,146	0.7	0.4	75.0
OPEC	4,725	3.0	0.6	400.0
TOTAL	159,571			

Sources: United States Department of Commerce, *Survey of Current Business,* Bureau of Economic Analysis (Washington, DC, October 1977, August 1978, and October 1984 issues). United States Department of Commerce, *News*, Bureau of Economic Analysis (Washington, DC, June 1985, BEA 85-33).

compared to its real market value, it is apparent that the turnaround in Canada's direct investment position, starting in 1975 and accelerating after 1979, is a major phenomenon worthy of detailed study.

Canada as one of a family of investors

From Table 8.3 it appears that Canada is only one of several key players in the United States and its importance has been declining over this period. Table 8.3 provides a clearer breakdown of the share of Canada's FDI in the United States, relative to other nations. Over the last ten years Canada's stock of FDI in the United States has doubled, from some 5 billion dollars in 1975 to over 14 billion in 1984. However, its percentage share of the total stock of FDI in the United States has fallen from 19.3 per cent to 8.8 per cent over this period. In these ten years Japan has grown from a small investor of 2.1 per cent to slightly exceed Canada's current stock of FDI in the United States. The largest individual stock of FDI belongs to the United Kingdom (24 per cent). Canada's share of the stock of FDI has been overtaken by the Japanese share — 9.3 versus 8.8. There have also been increases in the value of OPEC and Middle East FDI in the United States, both of which are now about 3 per cent of the total.

Intra-industry FDI

Canada's focus upon the United States is not a singular characteristic; rather Canadian FDI is reflective of investment patterns of the major trading nations. In the last five years especially, the United States has been a magnet for FDI; indeed, the United States is now a net importer of FDI, whereas in the 1970s it was the world's largest exporter of such capital.

This reversal of investment flows suggests that traditional concerns over the foreign ownership of Canada's economy need to be balanced by an awareness of the maturing of the Canadian economy to the point where two-way flows of direct investment have become the norm. Such cross flows of direct investment are now common in most of the world's major industrialised nations, such as the members of OECD. This intra-industry FDI indicates that MNEs are now spreading their activities across leading economies, including those of the United States and Canada (Erdilek, 1985).

It is apparent that the US market is the major attraction for Canadian investors. Many large Canadian MNEs need to ensure access to the United States in order to achieve an efficient scale in their operations. A secondary motive for the upsurge in Canadian investment activity in the United States is the desire of Canadian MNEs to restructure and diversify their operations in a strategic sense. The trend towards increasing direct investment in the United States was accelerated in 1980 when the combination of Canadian energy measures in the National Energy Program, coupled with the increased intervention of the Foreign Investment Review Agency (at that time), contributed to an unfavourable investment climate in Canada. This led to large divestments of US-owned energy firms in Canada and a record amount of some $11 billion of Canadian FDI in the United States that year. (This consists of both repatriations of existing foreign-owned firms and direct investment by Canadian-controlled MNEs.) The assessment of political risk in a nation is ultimately subjective and it could be argued that the perceptions of political risk were specific to the 1980–82 period. Since then, however, the recession has made a relatively greater impact upon Canada than the United States, so the latter nation has remained relatively more attractive for FDI. There has been a maturing of Canada's MNEs in recent years, and their specific activities are now examined.

THE FIRM-SPECIFIC ADVANTAGES OF CANADIAN MULTINATIONALS

This second section of the paper examines the manner in which Canada's largest MNEs penetrate the vital US market. The nature, structure and performance of the largest Canadian industrial MNEs are analysed and the special firm-specific advantages (FSAs), usually in marketing, of each of the MNEs is identified.[1]

Most of the largest Canadian MNEs have FSAs in the production, distribution and trading of resource-based products, or are in mature product lines. Indeed, only one of the MNEs possesses the knowledge or technologically-based FSAs of the typical US, European or Japanese MNEs. The Canadian FSAs are related to the country-specific advantage (CSA) of Canada in resources and occur in firms in mature industries, in which the FSA often turns out to be marketing related, rather than technologically based. These MNEs engage in FDI when exporting to the United States is blocked by significant environmental constraints (often in the form of non-tariff barriers such as US contingent protection). The reasons for FDI and the manner in which the CSAs are internalised by the Canadian MNEs are discussed in more detail in Rugman and McIlveen (1985). Here the focus is upon the implications for strategic management of the marketing FSAs of these MNEs in a world of increasing global competition. This research is based upon interviews of the strategic planners of these MNEs and analysis of company annual reports and other published information.

The theoretical background for this work comes from a combination of two areas of analysis of the corporate enterprise. First, the work of Rugman (1980b, 1981) on the theory of internalisation is used as a basis for identification of the FSAs of each MNE. In this theoretical work it has been shown that each MNE has internalised, i.e. secured property rights over, a special differential advantage. Frequently this is in the form of a knowledge advantage (based on R&D expenditures which have generated a technological edge), but it may also occur due to marketing advantages, as in the possession of a well-established and respected distribution network, or even in more intangible aspects of the skills of the company management.

The second strand of theory used is the work by Michael Porter (1980) on competitive analysis, which is readily applicable in an international dimension. Here his emphasis on entry and exit barriers and the analysis of competitive forces as they influence the strategic planning of the corporation are applied in a global context. In Porter's model the firm needs to assess the environment in which it operates, especially the industry or industries in which it competes. Competition in the industry depends on five competitive forces: rivalry among existing firms, the threat of new entrants, the threat of substitution, and the bargaining power of suppliers and buyers. The goal of competitive analysis is to assess the strength of such competitive forces in order to determine the best strategy to adopt. Insight as to the strength of each force is available through analysis of entry and exit barriers in the industry.

The key entry barriers are: scale economies whereby existing firms enjoy production and cost advantages over new entrants; product differentation as rivals must break the barrier of existing brand loyalties; huge capital requirements involved in entering a new industry; switching costs necessary to change suppliers; access to distribution channels where established firms already have control of the distributors; and government regulation which may bar entry or impose licensing requirements on a new firm. Exit barriers include: the existence of equipment which is of such a highly technical nature that it has low marketability; fixed costs associated with settlements of contractual arrangements with workers and low productivity once it is known that liquidation will take place; strategic barriers if the business is fundamental to the firm's strategy and image; informational barriers where the absence of clear and accurate information makes it impossible to assess performance; emotional barriers associated with managerial pride in the company and the fear of loss of status; and government which may prevent a firm from exiting in order to preserve jobs or for other social reasons.

Identification of the Canadian multinationals

The 16 largest Canadian-owned companies are identified in Table 8.4. The firms are derived from a set of the 24 largest Canadian-owned firms from the 1982 *Fortune* International 500. Of these 24 firms a Canadian MNE is defined as a firm with a foreign-operating subsidiary in at least one foreign country and a minimum foreign to total sales ratio (F/T) of 25 per cent. These criteria reduce the set to the group of 16 MNEs. Canadian Pacific, the largest industrial corporation in Canada, is deleted since it is a holding company. Instead one of its subsidiaries, AMCA International, is included.

The 16 Canadian MNEs are almost all resource based or are in mature sectors. The industrial mix is as follows: pulp and paper 4; mining and metal manufacturing, 3; beverages, 3; and six other single-industry categories. The special cases include: NOVA, a petroleum resource MNE; Massey-Ferguson, the farm machinery manufacturer; Moore, the world's largest producer of business forms; Genstar, a vertical-integrated construction materials and mining resource MNE; and AMCA, the steel-related equipment manufacturer specialising in resource extraction and processing equipment. The only true high-tech Canadian MNE is Northern Telecom. It is the second largest manufacturer of telecommunications equipment in North America and is widely considered to have the most technologically advanced telephone switching equipment available.

In terms of sales, the Canadian MNEs are smaller than their US or European counterparts. The average size (from Table 8.4, converted to US dollars) is $1.752 billion. The largest 50 US and European MNEs by contrast have average sales of $16 and $12.4 billion respectively (Rugman, 1983). The Canadian MNEs' financial performance as measured by the return on equity (ROE) over the last ten years is 12.8 per cent compared to 14.3 and 8.5 for US and European MNEs respectively. The ROE for European MNEs is biased downwards by the significant presence of state-owned enterprises (Rugman, 1983). The risk of these returns as proxied by one standard deviation is 5.3, 3.6 and 5.2 respectively for Canadian, US and European MNEs. In short, the Canadian MNEs earn comparable returns to US MNEs but at greater risk, while earning higher returns at the same risk level relative to European MNEs.

Table 8.4: The largest Canadian industrial multinationals

Firm	Average sales 1978–1982 (billions)	F/T	S/T	1973–1982 ROE	1973–1982 SD
Alcan	5.169	na[b]	77	11.5	7.7
Seagram	2.991	92	92	10.4	2.8
Massey-Ferguson	2.688	93	93	6.5	6.5
Noranda	2.578	60	28	13.1	8.6
Hiram Walker	2.565	na	47	11.7	2.5
Northern Telecom	2.214	61	48	14.9	5.8
MacMillan Bloedel	2.135	88[a]	39[a]	8.9	7.6
Moore	1.991	90	90	17.3	1.9
NOVA	1.990	na	34	12.1	2.2
Inco	1.636	82	42	9.7	7.6
Genstar	1.584	na	52	14.4	5.6
Domtar	1.568	29[a]	8[a]	12.6	7.5
Abitibi-Price	1.505	66	14	13.4	5.5
AMCA	1.403	na	78	15.8	2.8
Consolidated-Bathurst	1.323	54	20	16.6	7.7
Molson	1.235	na	27	15.5	2.1
Mean	2.161	72	49	12.8	5.3

Notes: a. 1981. b. Not available.
F/T is defined as the rate of foreign (F) to total (T) sales.
S/T is defined as the rate of sales by subsidiaries (S) to total sales. (The difference between F and S is exports (E) from the home country nation.)
ROE is the mean return on equity, i.e. the ratio of net income after tax and before extraordinary items divided by the average net worth (value of shareholders' equity).
SD means standard deviation.
Source: Corporate Annual Reports.

STRATEGIC MANAGEMENT OF THE MNEs AND PUBLIC POLICY

In today's world of increasing global competition, large US, European and Japanese MNEs compete aggressively for market share and profits in every corner of the world. Yet Canadian MNEs have been surprisingly successful global competitors despite the intensity of competition and the relatively small size of the open Canadian economy. Analysis of the largest Canadian MNEs suggests a variety of reasons for this success.

While government seems to assume that Canadian business must move into high-tech areas the underlying logic of market forces and comparative advantage points towards a different outcome. Of course, there are a few large firms such as Northern Telecom which can make the transition to high technology, but on balance most Canadian-owned firms need not move in this direction. Instead, Canadian managers are already realising that excellent international performance is possible through creation of a value-added chain in the harvesting, processing and marketing of resource-based product lines. The success of the value-added chain requires sophisticated and effective managers, knowledgeable about modern methods of strategic management.

Canada is a resource-abundant nation which can design strategies for international success based upon its comparative advantage rather than upon the panacea of technology. The focus upon technology is dangerous since it leads to simplistic recommendations for public policy towards trade, investment and industry. International competitiveness is not necessarily dependent upon the degree of technological sophistication of a nation's industries.

Canadian attitudes towards international competitiveness assume a close causal relationship between the degree of technological intensity in industry and the share of world exports. The inevitable policy implication is for Canada to increase its level of R&D, so that exports of high-tech products can expand to improve the current account balance. This either leads to more government involvement in the investment, productivity and output decisions of the firms, through a system of informal taxes and subsidies, or to a more formal degree of state support and intervention in the economy. The focus on high technology ties in with the need for an industrial strategy, informal or formal. Inevitably such government-directed strategies contain nationalistic and protectionist elements which go against market forces, comparative advantage, the discipline of global competition and the successful experience of Canada's own multinationals.

Successful MNEs need not be in the US, Japanese and European mould, i.e. with advantages in proprietary knowledge and the embodiment of high technology. The Canadian pulp and paper, mining and liquor MNEs are non-traditional, yet successful MNEs. Furthermore, Moore Corporation is an example of a Canadian firm which developed FSAs to complement high technology rather than to rely upon it, as did Northern Telecom. Competitive analysis can lead to strategies which foster the growth of Canadian MNEs, whereas participation in high-tech industries in, of and by itself need not guarantee success.

Canadian MNEs demonstrate that the FSA of the multinational

can be in marketing and experience. The efficient marketing of resource-based product lines is the primary strength of many Canadian MNEs. Seagram, Moore and Massey-Ferguson are examples of the critical importance of marketing and distribution. Each has an extensive distribution network which gives it a distinct advantage over its competitors. In Massey-Ferguson's case, it is one of the few advantages which the firm continues to enjoy. These relationships help to reduce the environmental costs, especially the political risk, which is part and parcel of any foreign involvement. Effective distribution networks, market knowledge and experience result in favourable barriers to entry and the reduction of competitive forces. Switching costs, product differentiation and control of the distribution channels are effective even when cost, scale and government barriers do not exist.

The FSAs of Canadian MNEs also build upon Canada's country-specific advantages (CSAs). The firms either own mineral deposits, have established long-term leases for timber rights, or own energy resources which are cheap and abundant relative to foreign rivals. In short, Canadian MNEs have internalised Canada's CSAs in resources, which in turn leads to special firm-specific advantages. The only high-tech Canadian MNE is Northern Telecom. FSAs which build upon CSAs can form formidable barriers to entry. Such FSAs give Canadian MNEs access to important sources of raw materials such as cheap Canadian hydroelectric power. Nationalism can also be an FSA since favoured Canadian firms may receive preferential access to resource and grants from the responsible governments. Canadian MNEs also benefit from links with provincial governments which reduce information costs and political risk.

In conclusion, the Canadian example illustrates that a significant high-technology presence is not necessary for successful MNEs. A more basic determinant of international competitiveness is the underlying comparative advantage of the nation. A country like Canada can build upon its country-specific advantage in resources. Management can then internalise the nation's country-specific advantages into sets of firm-specific advantages. These create a value-added chain of resource-based product lines just as profitable and socially efficient as high-tech industries.

NOTES

1. This is a summary of a larger study on the strategic management of Canadian multinationals, the research for which was supported by the Social Sciences and Humanities Research Council of Canada. Additional support was received from the Department of External Affairs of the Government of Canada and the Dalhousie Centre for International Business Studies. The author thanks John McIlveen, Research Associate at the Dalhousie Centre for International Business Studies in 1983–4, for his help on this project. A detailed discussion of this section appears in Rugman and McIlveen (1985).

9

The Timing, Mode and Terms of Technology Transfer: Some Recent Findings[1]

D.G. McFetridge

INTRODUCTION

As the variety of topics covered by the papers prepared for this conference suggests, international technology transfer is becoming increasingly important in terms of both the public policy interest it has stimulated and the productive research opportunities it offers.

This paper reports some recent empirical findings regarding two aspects of the technology transfer process. These are: (a) the speed of international technology transfer and (b) the choice of transfer mode. The findings are quite general but the research has been motivated by and is discussed within the context of the concerns of Canadian economic policy.

Findings related to the speed of international technology transfer are discussed in the first section. The speed of international technology transfer is often expressed in terms of the time lag between the introduction of a new technology and its initial transfer abroad. The Canadian concern is with both the time lag between the introduction of a new technology and its transfer to Canada and the order in which Canada receives new technologies from abroad.

Findings relating to the choice of transfer mode are reported in the second section. The emphasis here is on the crude distinction between arm's-length and intracorporate or internal transfers and on their relationships with the transfer order and transfer lag respectively. A modest contribution of this paper is to allow statistically for the joint determination of transfer mode and transfer order by the economic and technological characteristics of the transfer situation.

Of potentially greater interest is the investigation of various forms of internal and arm's-length transfer arrangements. As Davidson (1980) has shown, the multinational enterprise can be organised in a variety of ways, some of which are more conducive to technology transfer than others. Similarly, attention has recently turned to the task of explaining the incidence of various features of arm's-length technology transfer agreements (Caves, Crookell and Killing, 1982; Shapiro, 1985).

Some recent Canadian evidence in this area will be discussed in the concluding section of the paper.

THE TIMING OF INTERNATIONAL TECHNOLOGY TRANSFER

With respect to the timing of international technology transfer, there are a number of issues of interest. These include: (1) the change in the average transfer lag over time and (2) the magnitudes and sources of international differences in transfer lags.

Insofar as the behaviour of transfer lags over time is concerned, the general impression is that they have become much shorter. Evidence on this matter comes from a variety of sources.

In their study of the transfer of 406 new product innovations by 57 US-based multinationals, Vernon and Davidson (1979) found that the proportion of these innovations transferred abroad within one year of US introduction rose from approximately 10 per cent before 1955 to over 30 per cent after 1970 (see Table 9.1 for details).

Edwin Mansfield and his associates have also gathered a great deal of evidence on this issue. In a five-country (France, Germany, Japan, USA UK) study of the plastics, semiconductors and pharmaceuticals industries, these authors found that the imitation lag declined in the plastics industry but did not show a statistically significant decline in the pharmaceuticals and semiconductors industries over the 1950–70 period (1982, p. 35).

In a study of 65 technology transfers by 31 US-based multinationals, Mansfield *et al.* (1982, pp. 36–8) found that the proportion of technologies transferred to affiliates in developed countries within the first five years of US introduction increased from 27 per cent in the 1960–68 period to 75 per cent in the 1969–78 period. There was, however, no tendency for the lag with which technologies are transferred to affiliates in developing countries or to arm's-length licensees (or joint venture partners) to decline.

Some statistical evidence on the acceleration of the multinational technology transfer process can be derived from data from Davidson (1980) and the Multinational Enterprises Database. It takes the form of a regression of the age of a technology (in terms of years since

Table 9.1: Transfers abroad of 406 innovations and 548 imitations of 57 US-based multinationals classified by period of US introduction

	Percentage transferred abroad within 1 year of US introduction		Percentage tranferred abroad within 2–3 years of US introduction	
Period of US introduction	Innovations	Imitations	Innovations	Imitations
1946–50	11.4	4.3	15.2	2.9
1951–55	7.0	7.8	5.3	13.3
1956–60	16.0	11.4	21.3	15.2
1961–65	26.9	19.6	17.5	16.7
1966–70	28.2	30.9	17.2	14.9
1971–75	38.2	33.8	26.2	10.8

Source: Vernon and Davidson (1979, Tables 11, 12).

first foreign transfer) at the time of transfer on the number of prior foreign transfers (the transfer order). This relationship should take the form of a cubic with age initially increasing less than proportionately with the number of transfers and ultimately increasing more than proportionately. If the international diffusion process is being compressed this relationship should shift downward over time. That is, the time period during which a given number of foreign transfers are completed will be shorter for technologies introduced more recently.

The results reported in Table 9.2 provide fairly strong support for the hypotheses regarding both the age-order relationship and the acceleration of the transfer process. Other characteristics of the technology, including its importance and whether it is an input or final product technology, do not appear to affect the age-order relationship.

This acceleration of the rate of international technology transfer has been discussed by both Vernon and Mansfield in product cycle terms. According to the product cycle theory, new technologies are initially embodied in the exports of the innovating country. As the technology becomes better known it is transferred abroad, first to affiliates of the innovating firm and ultimately to arm's-length licensees.

There is an extensive product-cycle literature which attempts to link exports or trade balances to R&D expenditures. More recently a policy literature has emerged which argues on product cycle grounds that comparative advantage (or at least trade patterns) can be 'engineered'. The argument is that, by spending large amounts

Table 9.2: The relationship between age and transfer order

Independent variable	Coefficient (t-ratio) OLS	TOBIT[a]
const.	7.00	2.26
(PTR + 1)	2.70	3.83
	(14.88)	(19.87)
$(PTR+1)^2$	−0.19	−0.34
	(7.46)	(14.05)
$(PTR+1)^3$	0.005	0.009
	(5.00)	(11.11)
YR	−0.16	−0.16
	(10.95)	(10.38)
INPUT	−0.16	−0.19
	(0.65)	(0.82)
MAJOR	−0.01	0.07
	(0.55)	(0.35)
n	991	991
R^2	.56	.57

Variable definitions:

AGE = Year of the jth foreign transfer of the ith technology less the year of the first foreign transfer of the ith technology.
= dependent variable.

PTR = number of prior foreign transfers of the ith technology as of the jth transfer.

YR = year of US introduction of the ith technology.

INPUT = 1 if the ith technology is a capital good or an intermediate input, zero if it is a consumer good.

MAJOR = 1 if the ith technology is deemed to have had a major impact and if it was a novel product (rather than an adaptation), zero otherwise.

a. TOBIT coefficients are $dE(AGE)/dX = (dAGE/dX|AGE > 0)P(AGE > 0)$. The t-test is valid asymptotically in the case of TOBIT.

on R&D and encouraging (possibly by means of initial protection) domestic producers to move rapidly down the learning curve a nation can pre-empt world markets in high technology products.

The compression of the product cycle reduces this potential R&D-exports relationship. Mansfield (1984, pp. 136–8) makes this point persuasively. He reports that a 1974 survey of 23 US firms revealed that exports were envisaged as the principal mode of international technology transfer during the first five years after introduction in only 15 per cent of the R&D projects examined. For the overwhelming bulk of the projects (75 per cent) the initial international dissemination of the technology was to take the form of transfers of the technologies themselves to affiliates.

These findings raise serious questions regarding the pay-off from successful high-tech targeting strategies. The pay-off has often been alleged to take the form of the establishment of a high-wage domestic export sector. The possibility also exists, however, that successful technologies, developed at public expense, will simply be exploited in other countries.

This apparent compression of the product cycle has been observed largely in connection with US-based firms. Proponents of high-tech targeting would, with reason, argue that what is being observed here is not so much a compression of the product cycle as a secular decline in US manufacturing competitiveness. While this issue has been hotly debated the relevant point here is that it would be useful to have some information on the evolution of the product cycles of non-US-based firms.

Transfer or imitation lags can also be compared across countries at a given point in time. Among the major studies of international differences in imitation lags are those of Hufbauer (1966) (synthetic materials), Tilton (1971) (semiconductors) and Nasbeth and Ray (1974) (various industrial processes).

The major question in this area is, of course, whether there are systematic international differences in the speed with which new technologies are acquired. Specifically, are there systematic differences in the characteristics of early and late adopters?

This issue has been of particular concern in Canada. Canadian observers including Daly and Globerman (1976) and the Economic Council of Canada (1983) maintain that Canada lags behind other industrial countries both in the speed with which the first domestic adoption of new foreign technologies occurs and in the rate at which these technologies spread to other domestic firms.

Evidence on international differences in transfer lags during the period 1960–79 from three major databases has been assembled by McFetridge and Corvari (1985). This evidence is reproduced in Table 9.3. It is clear that when the mode of transfer is held constant the average transfer lag to Canada over this period has been at least as short as the average for Western European countries or developed countries taken as a group.

The transfer lags reported in Table 9.3 reflect the speed with which domestic manufacture of new products developed elsewhere begins. The concerns of Daly and Globerman and the Economic Council are more with the speed with which domestic firms acquire capital goods which embody new technologies (such as numerically-controlled machine tools) for use in their manufacturing activities. The two sets of evidence taken together imply that, while domestic

Table 9.3: Mean international transfer lag (1960–79)

Multinational Enterprise Database	Canada			Europe			Rest of the World		
	Intracorporate	Arm's-length	Both	Intracorporate	Arm's-length	Both	Intracorporate	Arm's-length	Both
Mean lag	6.93	10.0	7.11	10.27	10.86	10.42	11.01	12.40	11.70
Observations	115	7	122	340	116	456	323	233	556

Economic Council of Canada	Canada		
	Intracorporate	Arm's-length	Both
Mean lag	5.8	8.8	6.94
Observations	37	19	56

Mansfield and Romeo (1960–78)	Overseas developed countries[a]		Less developed countries
	Intracorporate	Licensing/joint venture	Intracorporate
Mean lag	5.8	13.1	9.8
Observations	27	26	12

a. Including Canada

Data sources: Multinational Enterprises Database (Vernon and Davidson, 1977), Economic Council Database (DeMelto *et al.*, 1982) and Mansfield and Romeo (1980, pp. 737–50).

Source: McFetridge and Corvari (1985).

manufacture of new products has been initiated in a timely fashion, the technology of manufacturing itself has been slower to adapt to innovations occurring abroad.

The data reported in Table 9.3 also confirms the relationship between the mode of transfer and the transfer lag. Transfer lags tend to be shorter in the case of intrafirm (internal) transfers. More will be said about this issue subsequently.

Another way of looking at international differences in transfer lags is to examine the order in which Canada and other countries receive new foreign (in this case US) technologies. Some information of this nature for Canada and France is contained in the Multinational Enterprises Database and is summarised in Table 9.4. The position of Canada in the transfer order has slipped steadily since the early 1950s while the position of France shows no discernible trend. The table implies that, prior to 1965, Canada was, on average, the second recipient of the new US technologies in the sample while France was, on average, the fourth. After 1965, the average position of both was that of third recipient.[2]

There have been a few attempts to determine whether the industrial characteristics of early adopting countries differ systematically from the characteristics of late adopters. Two early studies by Swann (1974) and Nasbeth and Ray (1974) found, first, that there was a trade-off between the timing of the first adoption and the domestic diffusion rate. Early first adoption is associated with a slower domestic diffusion rate. There is not necessarily a big pay-off for getting a technology into the country earlier.

Second, both the international transfer lag and the domestic diffusion rate depend to a considerable extent on such largely exogenous national characteristics as factor endowments (labour-saving innovations are adopted less quickly in low wage countries) and industrial composition (NC machine tools appeared earlier in countries with large aircraft industries and synethetic rubber appeared earlier in countries with large rubber tyre industries).

Third, both international transfer lags and domestic diffusion rates are thought to be shorter the more open is the national economy to foreign competition.[3] Further evidence on the effect of competition on national imitation lags comes from the work of Mansfield *et al.* (1982, p. 35), who find that greater domestic industrial concentration is associated with a shorter average national imitation lag in the case of plastics and a longer imitation lag in the case of pharmaceuticals. No statistically significant relationship was observed in the case of semiconductors. These authors also found that shorter

Table 9.4: National position in the transfer order: Canada and France by time period

Period	Canada Number of transfers	Canada Average position	France Number of transfers	France Average position
1946–50	2	2.00	2	3.50
1951–55	16	1.62	6	1.83
1956–60	27	1.67	8	3.75
1961–65	50	2.10	32	4.22
1966–70	32	2.56	28	3.32
1971–75	22	3.91	17	3.00
1976–78	2	3.50	4	4.50
Weighted mean				
1946–65		1.89		3.81
1965–78		3.12		3.31

Source: Multinational Enterprises Database.

imitation lags were associated with larger R&D expenditures in the semiconductors and plastics industries but not in the pharmaceuticals industry.

Fourth, transfer lags are also likely to depend on national policies toward foreign direct investment. Specifically, other things being equal, transfer lags will be longer in the case of countries which screen or limit foreign direct investment.

For reasons which will be discussed in detail in the next section, intracorporate transfer is usually less costly than arm's-length transfer for the newest and most radical technologies. The cost advantage of internal transfer diminishes as a technology becomes more widely used and understood. The prohibition or discouragement of the ownership linkage necessary to internalise a transfer may, therefore, result in a postponement of transfer until an arm's-length arrangement is feasible.

Some evidence on the relationship between the economic characteristics and policies of a country and the order in which it receives a given new technology can again be gleaned from the Multinational Enterprises Database. In the simplest terms the model is:

$$O_{ij} = f(C_j, T_i, P_j)$$

where

O_{ij} = the order in which the ith technology is received by the country.

C_j = economic characteristics of the jth country.

P_j = policies of the jth country with respect to foreign investment.

T_i = characteristics of the ith technology.

In general, the jth country should have an earlier position in the transfer order the larger is the market to which its domestic producers have access and the more costly it is to serve that market with imports. Indicators of market size include national GNP and GNP *per capita*. Indicators of the size of the market for the latest technologies might include the literacy rate and the *per capita* stock of such consumer durables as automobiles and television sets. The extent to which economic and technological sophistication advances the national position in the transfer order should be greater the more novel is the technology involved.

The cost advantage of serving the market of the jth country from a local source will be greater the higher are its (effective) tariff rates, the lower are local input prices and the lower is the cost of importing the requisite technology. As suggested above, a policy of detailed screening or of prohibiting foreign direct investment can increase the cost of technology transfer and thus relegate the country pursuing this policy to a lower position in the transfer order.

TOBIT estimates of equation (1) are reported in Table 9.5. They provide general confirmation of the hypotheses advanced above. In particular, nations with no screening *or* equity controls tend to be earlier in the transfer order than those which screen or maintain ownership controls. While this is interesting and accords with expectations, it would be unwise to read too much into it. A more detailed analysis reveals that there is apparently no difference in the transfer order between countries in which there are some controls and those in which controls are pervasive.

The size of the national economy, income *per capita* and the literacy rate all contribute, as expected, to the advancement of the national position in the transfer order. The relative importance of the manufacturing sector is also important where input technologies are concerned. Other things being equal, higher national tariff barriers tend to be associated with an earlier position in the transfer order. Other specifications of the model revealed that the national stocks of various consumer durables had no effect on the transfer order. Nor did the advent of the European Economic Community change the relative position of its members in the transfer order.[4]

In equation (3) in Table 9.5 the variable ENEQ which reflects the absence of screening and equity controls is replaced by a dummy

Table 9.5: National characteristics and policies and the international transfer order

Independent variable	Eq. 1 OLS coeff.	Eq. 2 TOBIT[a] coeff.	η	Eq. 3 TOBIT[a] coeff.	η
Const.	8.70	5.75		5.28	
GNPPC	−0.03	−0.03	−0.05	−0.04	−0.07
	(2.21)	(2.27)		(3.04)	
LIT	−0.05	−0.04	−1.46	−0.04	−1.33
	(6.21)	(5.44)		(4.95)	
DMFR	−0.009	−0.008	−0.07	−0.013	−0.12
	(1.35)	(1.38)		(2.27)	
GNPR	−0.26	−0.24	−0.72	−0.25	−0.77
	(6.21)	(4.25)		(4.59)	
DHBT	−0.76	−0.42	−0.04	0.11	0.01
	(2.33)	(1.44)		(0.50)	
T	0.09	0.10	0.81	0.12	1.02
	(5.60)	(6.12)		(7.64)	
ENEQ	−1.07	−0.87	−0.22	—	—
	(3.69)	(3.30)		—	
P	—	—		−1.15	−0.32
				(5.73)	
n	991	991		991	
R^2	.20	.21		.24	

a. Coefficients are $d\{E(PTR)\}/dX = (dPTR/dX|PTR > 0)P(PTR > 0)$
η is the elasticity of the expectation evaluated at the sample means.

Variable definitions:

PTR = number of prior foreign transfers of the ith technology at time of transfer to the jth receiving country
= dependent variable

GNPPC = GNP *per capita* of the jth receiving country

LIT = literacy rate of the jth receiving country

DMFR = MFR × INPUT

MFR = percentage of the national GNP of the jth receiving country accounted for by manufacturing.

INPUT = 1 if the ith technology is a capital good or intermediate input; zero otherwise.

GNPR = GNP fractile rank of the jth receiving country.

DHBT = 1 if the jth receiving country imposes relatively severe restrictions on imports; zero otherwise.

T = year of transfer of the ith technology to the jth country, 1945 = 0.

ENEQ = 1 if the jth receiving country neither screens foreign investment nor imposes ownership controls on foreigners, zero otherwise.

P = 1 if the transfer of the ith technology to the jth receiving country is internal.
= 0 if the transfer of the ith technology to the jth receiving country is arm's-length.

Sources: see Davidson (1980) and Davidson and McFetridge (1985).

variable, P, which is equal to one in the case of technology transfers between affiliated firms and zero in the case of arm's-length transfers. The results indicate that, given their economic characteristics, the countries to which intra-firm technology transfers were made tended to be earlier in the transfer order.[5]

The specification of equation (1) employed here is rudimentary. It neglects the influence of relative input costs. It employs national rather than technology-specific measures of tariff protection and market size. No allowance is made for the interaction of national and technology characteristics. All countries and all technologies are pooled even though each country in the sample receives a different set of technologies.

Despite these and other limitations, these initial results are intuitively plausible and imply that a more refined analysis would be productive. Among the refinements which might be attempted are the analysis of the effect of the characteristics of various technologies on the speed with which they are acquired by a single country and the analysis of the effect of national characteristics on the speed with which a single technology is acquired by various countries. The merits of pooling across technologies could then be investigated.

Of course, the timing of a technology transfer is not the only thing that matters. It does not matter how early a technology is acquired if all its rents are appropriated by its foreign owner. Rents may be shared with domestic nationals in a variety of ways including consumers' surplus, tax revenues to host governments and spillovers to domestic producers. Globerman (1979) found some evidence of spillover benefits accruing to Canadian-owned firms from foreign-owned firms in the same industry. More recently, Mansfield *et al.* (1982) found evidence of the same type of spillover benefits in the UK (pp. 46–7). They also found weak evidence that US transfers of technology abroad tended to hasten the access of non-US competitors to these technologies.[6]

THE CHOICE OF TRANSFER MODE

The choice of technology transfer mode is, in its simplest form, a choice between an intracorporate transfer and an arm's-length licensing arrangement. These are the polar cases of what is in fact a continuum of possible arrangements between the transferor and the recipient. Among the intermediate arrangements would be long-term technology transfer or sharing agreements and various forms of joint ventures.

Recent attempts to explain the circumstances under which a particular transfer mode might be chosen have met with some success. Drawing on the work of Williamson (1971, 1975, 1979) among others, economists have argued that an intra-firm transaction will be less costly than an arm's-length transaction in situations characterised by irreversible commitments (transaction-specific assets) and costly measurement (information asymmetries). These circumstances give rise to opportunism, the incentive for which can be attenuated by internalisation.

In their recent studies, Davidson and McFetridge (1984, 1985) hypothesise that the scope for opportunism, hence the benefits of internal transfer, would be greater the newer and more radical is a technology and the fewer the prior transfers of it. These hypotheses receive general statistical support when tested using data from the Multinational Enterprises Database.

Davidson and McFetridge also find that, given the characteristics of the technology and of the parties to the transfer there has been a trend away from internal transfer since about 1965. They attribute this finding, which has been confirmed by a number of other investigators, to an increase in competition in international markets for new technologies (p. 262, n. 10).

An increase in competition may reduce the relative cost of licensing for a number of reasons. The existence of competition reduces the scope for opportunism. In the simplest terms, the existence of competitors disciplines a transaction by increasing the ability of the parties involved to turn elsewhere in the event of disputes. The existence of competition also results in a faster bidding down of quasi-rents, reducing the potential gain from opportunistic behaviour. In both cases there is a reduction in the magnitude of the potential gain from internalisation.

Other reasons for the apparent increase in the importance of arm's-length transfers include a general increase in experience with technology transfer and a growing familiarity with potential technology source or recipient firms in other countries. Others such as Contractor (1983b) also cite the demands of some countries that technologies be transferred by licence rather than internally as another explanatory factor.

The discussion of the factors bearing on the choice of transfer mode suggests that the earlier transfers of the newest and most radical technologies are more likely to be internal. The discussion of transfer lags in the previous section concludes that internal transfers will tend to be earlier in the transfer order than arm's-length transfers. The implication is that the transfer lag (or order) and the transfer mode are determined simultaneously.

Put simply, the order in which the jth country receives the ith technology will depend on the transfer mode and the economic characteristics of the jth country. The transfer mode chosen will depend on the transfer order, the chracteristics of the technology and the policies of the jth country towards foreign direct investment.

This two equation model can be written in general functional form as:

$$O_{ij} = f(\hat{M}_{ij}, C_j) \quad (2)$$
$$M_{ij} = g(\hat{O}_{ij}, T_i, P_j) \quad (3)$$

where M_{ij} = the mode of transfer of the ith technology to the jth country.

Note that equation (1) in the previous section can be regarded as the reduced form of this model. The age-order equation reported in Table 9.2 could be added to this model. The model could be recursive with order and mode being determined jointly and order then determining age. It might also be fully simultaneous with age being determined by order but helping to determine mode. The approach taken here is to assume that the model is recursive.

Similar models have begun to appear in the literature. In her study of technology transfers to Canada, McMullen (1982) estimated a three-equation model in which the dependent variables were the transfer lag, the transfer mode and relative innovation cost. She finds that the transfer lag was shorter: (a) in the case of intracorporate transfers; (b) in the case of incremental innovations; and (c) in the case of smaller (less costly) innovations.

The results of the estimation of equations (2) and (3) simultaneously using data which are, again, drawn from the Multinational Enterprises Database are reported in Table 9.6.[7] As hypothesised, the mode of transfer chosen depends on the characteristics of the technology, including the number of times it has been transferred, the characteristics of the transfer and the policies of the government of the receiving country. The transfer order depends on national characteristics (interacting with the characteristics of the technology) and the transfer mode. Public policies which alter the transfer mode also affect the national position in the transfer order.

Table 9.6: Estimates of a simultaneous transfer mode and order model

$$1)\ PTR_{ij} = 5.65 - \underset{(4.28)}{2.54\hat{P}_{ij}} - \underset{(2.99)}{0.06GNPPC_j} - \underset{(4.07)}{0.03LIT_j} - \underset{(2.99)}{0.02\,DMFR_j} - \underset{(4.50)}{0.25GNPR} - \underset{(0.23)}{0.05\,DHBT} + \underset{(7.87)}{0.14T}$$

$$2)\ P_{ij} = -1.18 - \underset{(3.59)}{0.08\,\hat{PTR}_{ij}} - \underset{(3.24)}{0.04USAG_i} + \underset{(0.87)}{0.08IMIT_i} + \underset{(6.80)}{0.67SIND_i} + \underset{(3.87)}{0.01PRND_i} - \underset{(3.16)}{0.17LNPV_i} + \underset{(3.37)}{0.33ENEQ_j} + \underset{(2.88)}{0.10T_{ij}} - \underset{(1.04)}{0.001T_{ij}^2}$$

n = 991

Variable definitions:
Equation (1) see Table 9.5.

USAG = time lag between US introduction of the ith technology and its first foreign transfer.
$IMIT_i$ = 1 if the ith technology is an imitation, zero otherwise.
$SIND_i$ = 1 if the ith technology is in the same three digit SIC industry as the transferor's principal line of business, zero otherwise.
$PRND_i$ = R&D intensity of the transferor.
LNPV = natural logarithm of the number of prior international technology transfers made by the transferor.

CONCLUSIONS AND SUGGESTIONS FOR FUTURE RESEARCH

The statistical results reported in this paper lead to the, perhaps obvious, conclusion that the mode and timing of technology transfer: (a) have changed over time in the direction of shorter transfer lags and proportionately more arm's-length transfers; (b) are jointly determined; and (c) are also determined by the characteristics of the economic and technological environment and by public policy.

The screening or limitation of foreign direct investment will result in the postponement of some and the elimination of other technology transfers. At the same time it may increase the domestic spillover benefits of infra-marginal transfers. Whether there is a net gain from screening or ownership controls is an open question although there are several reasons (see McFetridge, 1985) to believe that this is unlikely. The recent reduction in the scope of the screening activities of the Foreign Investment Review Agency (now called Investment Canada) would, therefore, appear to have been a constructive policy change.

As far as future research is concerned there are significant benefits from a more detailed comparison of transfer modes. As suggested at the outset, there are a number of ways in which a multinational enterprise can be structured, some of which will be more conducive to technology transfer than others. Governments considering the imposition of a particular corporate form, such as world product mandating, on foreign-owned firms should be aware of the consequences of that form for technology and information flows. Some research of this nature has been conducted by Bishop and Crookell (1985).

There are also a number of mixed or quasi-internal transfer arrangements, most notably joint ventures, the characteristics and comparative advantages of which are beginning to be investigated. Similarly, as Williamson (1979) has suggested, there is a continuum of so-called arm's-length arrangements. There is no systematic evidence as yet on the respective comparative advantages of various forms of arm's-length arrangements.

Studies of arm's-length technology transfer arrangements have historically concerned themselves with the restrictions on domestic licensees which these agreements have often involved. The extent of these restrictions has been investigated in a Canadian context by Killing (1975), Caves, Crookell and Killing (1982) and DeMelto *et al.* (1980).

In each study the incidence of market restrictions, grantbacks and tying arrangements is measured. To date litle progress has been made in explaining the relationship between the circumstances of the technology transfer and the incidence of various restrictive clauses or combinations thereof in licensing arrangements. The determination of the nature of this relationship is important for policy purposes. Restrictive clauses are functional in that they are preferable, from the viewpoint of the licensee, to alternatives such as higher royalties, larger front-end payments or no transaction. It would be productive to establish the nature of the likely response to a policy of prohibiting certain sets of restrictive clauses.

As far as the timing and ordering of the technology transfer is concerned, much remains to be done. One study of interest would be to distinguish the respective characteristics of technologies which a particular country, such as Canada, receives early, late or not at all. Another would be to distinguish the respective characteristics of countries which receive a particular technology early, late or not at all.

NOTES

1. This paper is a revised version of the paper prepared for the Round Table. The research assistance of Ronald Corvari is gratefully acknowledged. Thanks go also to Bill Davidson who has generously allowed me to make use of his data.

2. Davidson (1982) draws the same conclusion for Canada. McFetridge and Corvari (1985) compare the mean order of first and subsequent transfers to Canada and Western Europe. The average position of Canada is found to have been declining since the early 1950s. The average position of Western Europe as a whole has been declining since the mid-1960s.

3. It can also be argued, however, that as far as new product technologies are concerned, given market size, higher trade barriers may be associated with earlier technology transfer. See Swann, (1974, p. 63), Nasbeth and Ray (1974, p. 312) and Daly and Globerman (1976, p. 312).

4. In an earlier version of this paper, the specification of model (1) included a dummy variable for Canada. When the barriers to trade dummy was introduced the Canada dummy became insignificant. The sample size is also smaller by 115 observations than in the earlier version. This is the result of the elimination of technologies for which the year of the first foreign transfer could not be inferred.

5. For purposes of this analysis affiliation involves an equity interest of 95 per cent or more. An arm's-length relationship involves an equity interest of 5 per cent or less. Relationships involving equity interests between 5 and 95 per cent are not examined.

6. Access to new process technologies was accelerated by an average of 3.1 years and access to new product technologies was advanced by 0.4 years. These acceleration estimates were not statistically significant (pp. 39–40).

7. Equation (1) is estimated by the TOBIT method and reported as equation 1 in Table 9.6. Equation (3) is estimated by the probit method and is reported as equation 2 in Table 9.6. A two-stage estimation method suggested by Maddala (1983, pp. 242–7) is used. The interpretation of probit coefficients is explained in Maddala, p. 23. TOBIT coefficients are as defined in Tables 9.2 and 9.5. The author is not aware of an accepted goodness-of-fit measure for a simultaneous limited dependent variable system. The mode equation is significant at the 0.5 per cent level using a single equation likelihood ratio test.

Part Three: Industry Studies

10

The Protection of Intellectual Property: Pharmaceutical Products in Canada

H.C. Eastman

The structure of the economies of scale and of the costs of firms in the pharmaceutical industry provide a strong inducement to the multinationalisation of operations and consequently to some international transfer of technology. The principal element of policy that permits firms to benefit from their international operations is the patent protection offered to their inventions abroad. Such patent protection also affects the behaviour of the firms including expenditures on research and promotion, pricing and the resulting profits. In establishing appropriate patent protection for a chiefly foreign-owned pharmaceutical industry which does most of its research abroad, the authorities of countries such as Canada face two problems — that of balancing the extent to which they wish to induce price as against non-price competition and that of deciding to what extent Canadian consumers should contribute through high prices to the worldwide profitability of the research-based firms.

This paper first raises some general considerations about the optimal rate of investment in the discovery and development of new products or processes, structural obstacles to reaching that rate and the efficiency with which research and development (R&D) is carried on. The paper then goes on to discuss the cost structure of the international pharmaceutical industry as it is relevant to innovation and to the international transfer of technology. It then describes the Canadian Patent Act and the pharmaceutical industry in Canada. The last section of the paper describes and evaluates the protection offered to intellectual property in the pharmaceutical industry by provisions for compulsory licensing in the Canadian Patent Act.

INVESTMENT IN INNOVATION

At a particular time and state of knowledge, it is possible to conceive of a number of possible research projects in a given field ranked by decreasing degree of promise in the eyes of a firm or of a number of firms in an industry. The promise is a function of the economic importance of the potential discovery and of the probability of success. If there is full appropriability of the results of invention, which is to say that the firm collects the full marginal product from its investment in each project, and there is only one firm capable of undertaking innovative activities in that field, the firm will push its R&D to the point where the marginal return is equal to the marginal cost of inventing. If the firm has foreseen correctly, it will thereby make monopoly profits.

If there is monopoly in invention in the sense that many firms engage in investment in invention, but there is no duplication (each firm chooses a different project), the same result obtains except that different firms make different monopoly profits, which are the difference between the cost and the marginal product. The firm with the marginal project makes only a competitive rate of return.

Under these assumptions of appropriability and no duplication the inventions industry provides the socially optimal amount of research.

OBSTACLES TO SOCIAL EFFICIENCY IN RESEARCH AND DEVELOPMENT

The organisation of R&D may be inefficient from a social standpoint for two broad classes of reasons. Firstly, duplication of research undertakings and analogous difficulties may lead to an inappropriate volume of research. Secondly, appropriability may be imperfect owing to the possibility of easy imitation of new proudcts or processes. This latter difficulty arises from the competitive and cost structure of the industry that produces the new product or uses the new process, not from those of the innovating activities themselves.

Firms use resources in a socially inappropriate way if they duplicate each other's research, because duplication constitutes overinvestment in particular projects and resulting waste in the sense that the marginal product of expenditures is lower than it would be under monopoly on the projects that are undertaken. The rate of return for the successful inventor is higher than the competitive rate except for the marginal project. Competing firms have an incentive to invest in invention so as to discover products or processes earlier than their rivals in a race to the patent office or to establish their brand and product in the minds of consumers. In the case in which duplication is possible, and in which the prize to the winner of the race and the probability of success are high, under competitive pressure all the firms in the industry together may well engage in so much research that the monopoly profit is absorbed by these unnecessary costs of research.

The second main reason for an inappropriate level of investment in innovation arises because of imperfections in the market for the resulting product or process and not in the competitive structure of the innovating activity itself. This arises where the product or process market is competitive and imitation is so easy that the innovator cannot recover the cost of the R&D.

A social objective for the process of innovation is that the innovating activity should yield a competitive return on the investment required in R&D. But the limitation on imitation that ensures such a return should not lead to monopolistic behaviour that limits consumption through high prices and limited quantities for consumers in excess of the amount necessary to provide that return. That is, consumers should be able to benefit as much from the invention as is consistent with an adequate return for the innovator.

Inventions vary in their potential profitability. The size of the market for the product incorporating the invention may be large or small. The importance of innovations also vary according to the cost reductions that are achieved if they are new processes or to their attractiveness and novelty if they are new products. The elasticity of demand for the product is also a factor, because a low elasticity permits high prices and profits.

The extent to which the innovating firm can appropriate the fruit of its investments in R&D varies widely between industries. It depends on the total height of all barriers to entry into the industry including the ease of imitation of the new product or process resulting from R&D. This is itself a function of whether the technical progress is incremental and slow, and of the extent to which the innovation is integrated in the production process of the firm.

Authorities on innovation widely agree that some industries are especially vulnerable to the imitation of their new products. Most exposed among these are chemical and especially pharmaceutical products. Hence the importance of patent protection for firms in these industries.

It is evident from the variety of competitive conditions in both innovating activities and the final product market and from differences in the appropriability of innovations by the innovating firms that an optimal degree of patent protection would vary depending on the characteristics of each invention. Patent Acts have historically attempted to make some allowance for the particular characteristics of a few industries, notably for pharmaceutical products, as is the case today in Canada and the United States and has been until recently in the United Kingdom, France and Italy.

THE STRUCTURE OF THE PHARMACEUTICAL INDUSTRY

A large part of the cost of pharmaceutical products is in research. Research costs are on average about 10 per cent of sales on a worldwide basis for firms that are active in Canada. This research is used both to discover a new product and to test its characteristics before it appears on the market. Once invented, most pharmaceutical products can be produced at low cost and are easily imitated. In consequence, some patent protection for pharmaceutical products to act as a barrier to entry is clearly required in order to induce the appropriate amount of research in discovery. The crucial question is how much patent protection is warranted.

The high front-end costs of research and clearance to obtain authorisation to market a product, combined with the low marginal costs of actually producing most pharmaceutical products, are what give the industry its international character under presently prevailing patent conditions. Once discovered, more of a given product can be produced at very low cost. Each unit brings a return to the manufacturer. The bigger the market, the greater the profitability of that product. Hence the incentive to extend the sale of each product to as many national markets as possible.

Patenting necessarily creates product differentiation and delay in the appearance of competitive products. Patenting prevents competitors from producing identical products and competition must be less direct. Two pharmaceutical products may have the same therapeutic use, but they must be differentiated in composition even if their differences are slight from the standpoint of therapeutic effectiveness, as is often the case. The differentiation of products required by patents is the basis on which a patentee promotes the sale of his product by informing potential consumers and their agents of its effectiveness and establishes his trade name and his firm's identity in their consciousness. Trade name and firm preference constitute in themselves barriers to entry and to competition additional to patent protection.

An effect of the research necessary to differentiate patented products is that, when a new drug appears on the market and proves to be profitable, potential competitors engage in R&D activities to produce a similar, but not identical, drug that will compete in the same market. The process of developing drugs that are similar in their composition or purpose is expensive and time is required both to develop and to obtain authorisation for marketing. In other words, the patent system creates a delay in the appearance of competition and so permits the first firm with a new drug to initially set high prices.

Patents and product differentiation are not the only impediments to the entry of new competitors into the pharmaceutical industry. Research and development, including the process of testing new drugs for toxicity and therapeutic effectiveness and taking them for approval through the regulatory process, is an expensive undertaking that can only be carried out effectively by large firms. The question of whether a research laboratory can be productive if it is small, say with 75 employees, or need be much larger is a debated question, but is not an issue here. The point is that developing a new drug and putting it on the market is very expensive and that the risk of failure in the development of a particular drug makes it advantageous from the standpoint of avoiding risk to carry forward a number of projects simultaneously. Thus the economies of scale for this part of the activities of the pharmaceutical industry are important and create a barrier to the entry of small potential competitors.

On the other hand, the manufacture of finished pharmaceutical products, which involves the blending, mixing, encapsulation, compounding and other processing of active ingredients and other components, can be carried out in small plants which typically produce a number of different products. Thus economies of scale do not exist as a barrier to the entry of firms into this stage of manufacturing.

It is otherwise with the production of fine chemicals and their synthesis into active ingredients. The average cost of producing active ingredients typically declines over very large outputs so that an efficient plant often supplies a substantial portion of the world market for that ingredient. Nevertheless, these declining costs do not constitute important barriers to entry in the manufacture of components that usually form 25 per cent or much less of the total cost of the final product.

The existence of patents leads to the integration of these three aspects of production in the pharmaceutical industry under the control of a single firm. The patent usually resides in the active ingredient, not the finished product. The patentee often carries out himself all three steps in the production of a finished pharmaceutical product. A typical integrated company carries out R&D, obtains patents worldwide, complies with the clearance procedures in many countries, produces the active ingredients in one or a few favourable locations in the world, and manufactures the finished products in many plants in the countries that constitute its major markets. The economies of scale that result from the high sunk cost of discovering a new drug combined with the low cost of manufacturing are the forces that lead to the characteristic multinationalisation of firms with patents on major drugs in this industry.

The mature research-based firms in the industry are typically multinational enterprises (MNEs). Such firm structure does not necessarily result in the transfer of technology to host countries, because the part of the firm's production that takes place in the host countries usually consists of manufacturing the dosage forms, which is not a technically sophisticated operation for most drug products. The production of active ingredients is limited to only a few locations in the world and is in any event chiefly a mass production chemical-flow process which is highly capital intensive and has little technological fall-out. It is only when the firm chooses to carry on either basic industrial research or clinical research that significant technological or scientific activity is stimulated in the host countries by the MNE.

Some firms carry on a small portion of their basic research in a few centres outside the six countries in which pharmaceutical research is concentrated. To that extent some technological transfer occurs. Most firms fund some clinical research in host countries either because they are required to do so by the health authorities as a condition of drug clearance for marketing or to take advantage of the presence of qualified clinical researchers or to familiarize the local medical community with the new products. Such funding supports research by providing funds and products to test. However, most of such clinical research in host countries consists disproportionately of testing on large groups of patients which is medically and scientifically less significant than more basic clinical trials.

Pharmaceutical firms also support medical and pharmaceutical research in the universities of host countries by means of research grants and provide physicians with opportunities to augment their knowledge by sponsoring conferences, seminars and similar activities. It seems then that the transfer of technology to most host countries is chiefly at the clinical level.

The economies of scale in the R&D stage and the product differentiation arising from heavy brand promotion constitute the chief barriers to entry into the industry. The firm may license another firm to manufacture and sell its products in markets in which it is not active. These are usually foreign markets. Royalties vary, but many appear to be about 10 per cent of the final selling price. Licensing arrangements frequently require the licensee to purchase the active ingredient from the patentee and the price may exceed the lowest price available.

During the past several years about 15 to 20 new drugs have been produced each year.[1] These new drugs are single chemical entities or synthesised drugs not previously available. Not all new drugs bring major therapeutic gains or acquire a large volume of sales. The Federal Drug Administration of the USA classifies new drugs according to its view of their therapeutic merit. Of the 60 new chemical entities introduced into the USA in 1983 and 1984, six (10 per cent) were judged to offer significant therapeutic gains, 13 (22 per cent) to offer modest therapeutic gains, and 41 (68 per cent) to have little or no therapeutic advantages over existing remedies. Drugs drawn from the two larger categories are those that are especially profitable. Moreover, many do not become major commercial successes because the incidence of disease for which they are used is low. About five new drugs each year turn out to be money makers. These drugs are introduced internationally with heavy costs of promotion. They then attract competition by imitators who invent around the patent so as to produce a drug that is similar but sufficiently differentiated from the original to avoid infringement suits.

The principal effect of imitation drugs is to take a share of the market and to limit prices by offering competition in the market place for products that are close substitutes. Consumers benefit when prices are reduced in addition to the improvement in health or comfort brought by the innovation. However, this competition is achieved at the cost of the resources absorbed in developing and clearing the imitative drugs for marketing and these costs plus those incurred in promoting the differentiated product limit the reduction in prices that is possible.

THE CANADIAN PATENT ACT

The governments of most countries protect inventions by granting patents which give the inventor the right to the exclusive use of an invention for a specified period of time. A country's patent provisions may protect the article or substance that is invented (a product patent), the process or processes by which the article or substance is made (a process patent), or the process and the product made by that process (a product-by-process patent), or any combination of the three. Different types of patent protection may be given for different types of invention within a single patent system.

Canada's patent protection with respect to medicine does not allow a patent to be held for the chemical substance but only for the process and the product by the process. Patents are granted for pharmaceuticals in Canada for the way the compound is made, not for the chemical compound itself. Patent protection is limited to process or product-by-process in many countries, but product patents are the rule in an increasing number of countries including the USA, the EEC and Japan. The increasing uniformity of patent laws is largely the result of the movement toward the harmonisation of policies in the EEC as an objective in itself and not of the judgement of individual governments as to optimal protection for particular products.

The 'width' or 'breadth' of protection which is granted by a patent may vary. During the history of the Canadian patent system, policy and judicial interpretation of the legislation have brought variations in how broad or narrow a patent could be. In the past, patent claims were interpreted rather narrowly. Patent protection was in general limited to what the applicant disclosed he had actually done. More recent judicial interpretation has allowed claims to be broader. The patent can protect the class of things which can reasonably be predicted from the discovery even if it was not achieved by the inventor at the time of application for the patent.

The period of patent protection differs internationally. Canadian and US patents last 17 years from the date of issue. Grants run from the date of filing for a period of 20 years in the EEC, as in Japan.

Exclusivity is not always absolute. In Canada, provision is made for non-exclusivity by various provisions of the Patent Act. Section 67 of the Act allows the Commissioner of Patents to revoke a patent or to grant a licence to a party not holding it to use the patent holder's process if he is satisfied the patent is being abused. Abuse consists of failure to 'work the patent', that is, to manufacture in

Canada or charging excessive prices. Similar provisions are found in the patent acts of all major countries except the USA, but they are generally of little practical significance.

Section 67 has been little used. Only 11 have ever been granted, none in the area of pharmaceuticals. By and large, if the patent holder does not manufacture in Canada, it is because cost conditions do not warrant it for him or for a potential licensee. Alleged abuse of the patent on the basis of price alone have led to no grants of compulsory licences under Section 67. In any event, that Section does not provide for compulsory licensing to import, which would be the effective remedy owing to the large economies of scale in producing active ingredients which do not justify production in Canada.

Use of similar provisions is also very low in other countries. There are provisions in some countries for compulsory licences to be granted in the public interest, but again their use has been extremely limited.

The nature of the policies specific to the pharmaceutical industry has varied by country and over time. The British Patent Act was amended in 1919 in response to concern about the lack of domestically-owned pharmaceutical firms in the UK, which was attributed to excessively broad product patent protection for foreign firms. It restricted the patent protection given to food and drugs to process or product-by-process, not to the product itself. The amendment also introduced compulsory licensing of patents to manufacture and to import in order to permit the entry of new firms. This legislation was widely imitated in the British Empire. It was introduced in Canada in 1923 in a form that required manufacture of the patented active ingredient in Canada. This provision of the Patent Act had little effect in Canada for many years.

Under the chairmanship of Senator Kefauver, the US Senate Subcommittee on Anti-trust and Monopoly investigated the behaviour and profitability of the pharmaceutical industry in the USA. The impression was widely disseminated that the industry set high prices, incurred excessive selling costs and at times disregarded the interest of the public in its successful quest for allegedly excessive profits. These proceedings were followed by three major inquiries in Canada during the 1960s each of which concluded that Canadian drug prices were too high in comparison with those of other countries and that the Patent Act should be amended.

In 1969 the Patent Act was amended by the introduction of Section 41(4) to permit the issuance of compulsory licences to import. Under this provision the Commissioner of Patents issues compulsory licences unless he finds good reason not to do so, which is rarely the case. He has the responsibility for setting payments of royalties that are designed to compensate the patent holder for costs of research 'leading to the invention'. He has set these at '. . . 4% of the net selling price of the drug in its final dosage form or forms to purchasers at arm's-length'.

Compulsory licensing to import pharmaceutical products is today a specifically Canadian mechanism for mitigating the monopoly power of firms and reducing prices for the benefit of consumers and tax-payers. However, national or regional jurisdictions in nearly all other countries have also instituted policies that are directed specifically at pharmaceutical products. Direct control of drug prices, limits on profits of pharmaceutical firms, restricted lists of drugs eligible for reimbursement in drug insurance schemes, maximum prices for reimbursement, limits on promotion expenditures and weakened patent protection are devices implemented singly or in combination virtually universally. The prevalence of such measures reflects a shared judgement that protection from price competition in the industry is excessive and leads to prices and to costs or profits that are unjustified. Patent protection is the principal barrier to price competition in this industry.

THE STRUCTURE OF THE CANADIAN PHARMACEUTICAL INDUSTRY

The pharmaceutical industry in Canada is composed of approximately 160 firms of which 40 per cent with 20 per cent of value of sales are owned by Canadians. These are chiefly small firms producing drugs that are off patent or unpatentable and a few drugs under voluntary licence. Thus the dominant firms are foreign-owned multinational patent-holding firms. Three of the five largest Canadian-owned firms chiefly engage relatively heavily in research, produce biological products and are supported by government. The two largest Canadian-owned firms derive a substantial portion of their business from the sale of compulsorily-licensed drugs the patents for which are held by MNEs. Of the four important firms producing compulsorily-licensed drugs, two are Canadian owned and two are owned in the USA, the Canadian-owned firms having by far the largest share of the production of such drugs.

Sales of the 70 compulsorily-licensed drugs in Canada amounted to $328 million out of a total of $1.6 billion for all ethical drugs or

20 per cent of total sales. The generic firms, which are those that hold compulsory licences, have not supplanted the patent-holding firms in the market for licensed drugs. Indeed, generic firms sold and paid royalties on only 32 of the 70 drugs on which compulsory licences had been issued. Their sales of these drugs were $46 million or 21 per cent by value of total sales of $217 million of these compulsorily-licensed drugs, the remaining 79 per cent being accounted for by the patent-holding firms' brand name products. The 21 per cent generic share translates to approximately 34 per cent by volume of the market in compulsorily-licensed drugs when one takes into account that prices charged by generic firms are half those of patentees. The sales of such drugs by generic firms amounted to 3 per cent of the sale of all pharmaceutical products in Canada. Generic firms sold another 14 compulsorily-licensed drugs, but did not pay royalties, presumably because the patents had lapsed. The 24 other patented products on which compulsory licences had been issued by 1983 had sales of $111 million by patent-holding firms, but none yet by generic firms.

Generic firms sell drugs other than those that are under compulsory licence. Their sales of all pharmaceutical products are about 8 per cent of the value of total pharmaceutical sales in Canada.

Generic firms have been more active in some therapeutic categories than in others. In 1983, they held 13 per cent of sales of anti-infective agents and from 6 to 9 per cent of the sales in five other of the 19 therapeutic classes according to the Commission's survey of the biggest firms in Canada.

The generic firms have introduced an element of vigorous competition in the market for pharmaceutical products in Canada. They have concentrated on selling to hospitals and pharmacies and have used price competition as their strategy. In 1983, the prices of generic drugs were 51 per cent of the prices of the patent-holding firms for substitutable brands. The consequence of compulsory licensing is that Canadian consumers and taxpayers paid $210 million less in 1983 than they would have done for the same drugs in its absence. The $210 million in estimated savings is the difference between the actual purchases by both pharmacies and hospitals of the 32 compulsorily-licensed drugs sold by both patent-holding and generic firms and the cost of those purchases if their prices had had the same relationship to US prices as did those of unlicensed drugs. It is thus a definite figure.

In comparison, the competitive strategy followed by patent-holding firms in Canada, as abroad, is to introduce on the market new products which may have entirely new indications or significantly improved effectiveness or which may be similar to existing successful products and are introduced in order to share these markets. The patent-holding firms also incur heavy promotion expenditures, directed in large part to physicians. The rivalry between these firms results in the introduction of new products and in promotion expenditures and not in price competition, the effectiveness of which is attentuated by the differentiation of products. In Canada during the past five years the weighted average of promotion costs to sales for the 55 major firms in the pharmaceutical industry has been 21 per cent whereas the ratio of research and development expenditures to sales was 4.5 per cent and of profits to sales 15 per cent.

The four largest firms account for nearly one-quarter of total sales of ethical drugs. The 12 largest firms account for half that market and the 30 largest firms sell over 80 per cent of the total. Most firms rely on a small number of products for the vast bulk of their sales. Data for 1982 indicate that the leading product of a particular firm accounts usually for betwen 25 and 35 per cent of its sales. The lowest percentage amongst the 45 leading firms accounted for by the four leading products was 21 per cent, the highest was 96 per cent. In general the four leading products accounted for between 30 and 40 per cent of an individual firm's sales. This means that the level of concentration by therapeutic class is much higher than it is for the overall market in ethical drugs. The four-firm concentration ratio exceeded 80 per cent of the combined market in five of 14 therapeutic classes. In several other classes the concentration ratio was close to 50 per cent.

The dependence of firm sales on a few leading drugs also explains the exceptionally high degree of instability in the ranking of firms by sales, when compared to other industries, for all except the three leading firms.

COMPULSORY LICENSING OF PHARMACEUTICAL PRODUCTS IN CANADA

The purpose of amending the Patent Act in 1969 was to lower the price of drugs in Canada. In 1983 the average price of drugs that were not compulsorily licensed in Canada, weighted by Canadian consumption, was about 80 per cent of the average price prevailing in the United States, a proportion that has varied little over the years.

The average price of compulsorily-licensed drugs sold by both the patent-holding and the compulsorily-licensed firms in Canada was approximately half the prices prevailing in the United States for the same drugs. The prices of licensed drugs sold by the compulsory licensees were half those of the patent-holding firms and also exerted downward pressure on the latter's prices. Thus compulsory licensing to import is exerting an influence in the intended direction.

However, patents are designed to strike a balance between assuring an adequate return to the inventing firm and extending the benefit of technological progress to consumers. Does compulsory licensing meet the former objective?

The effectiveness of compulsory licensing in maintaining a balance between assuring an adequate return to the inventor and extending the benefit of the invention to the consumer depends on the royalty that is set for the license. In the case of pharmaceutical licences in Canada, the Commissioner of Patents has interpreted the legislation requiring a royalty that recognises the 'costs of research leading to the invention' as addressing only the costs preceding the application for a patent and not the costs of development and testing necessary to bring the product to market. The latter costs often substantially exceed the former. His interpretation has been confirmed by the courts. The resulting 4 per cent royalty does not cover the costs of research actually incurred on their worldwide operations by MNEs operating in Canada, costs which appear to be closer to 10 per cent of their total sales. Hence if compulsory licensing on a world scale were initiated at the 4 per cent royalty rate, research would be discouraged.

However, the inadequacy of the rate of royalty for compulsory licences in a particular instance does not negate the logic of a system of compulsory licensing. A rate of royalty that reflected the actual costs of useful research would introduce into patent administration a desirable flexibility to reflect particular situations and needs. An adequate royalty would etablish a minimum cost for imitators that would reflect an adequate rate of return. Small inventions would in consequence not often become the subject of application for compulsory licenses, because many scarcely cover their costs. But, in the case of major innovations, where the prospects for profit from patent monopoly is very large, the correct royalty for compulsory licences would introduce a desirable element of competition that would keep down the rate of return of the innovator to a level adequate to reward him for the effort of innovation without granting him monopoly profits to the detriment of the consumer.

The further benefit of issuing compulsory licences at adequate royalty rates is that it discourages imitative invention of commercially important inventions. Without compulsory licensing, the high prices and profits of major new drugs induce other firms to engage in research to imitate a new drug, differentiating their own new brand sufficiently to avoid patent infringement. This form of competition, which is typical among patent-holding multinational pharmaceutical firms, does not result in much lower prices; instead, firms incur heavy promotion costs to promote their brand in addition to the research costs of inventing around the original patent. It is better to introduce competition with a compulsory licence, because it avoids the waste of resources used in imitating the successful product and in promoting the imitation. Moreover, Canadian experience confirms that firms holding compulsory licences compete on the basis of price, which results in greatly reduced prices and costs for consumers.

Where consumers' interests are protected by the possibility of compulsory licences for major new drugs, which have a potential for excessive monopoly profits under normal patent terms, patents should be broad so as to discourage imitative research.

It should be noted that compulsory licensing as it exists in Canada is entirely consistent with the Paris Convention which does not require common patent provisions across member countries, but only the extension of national treatment to foreign firms.

NOTES

1. Data in this paper are drawn from Commission of Inquiry on the Pharmaceutical Industry, *Report* (Ottawa: Supply and Services Canada, 1985), *passim*.

11

International Investment, Protection and Technical Transfer: a Preliminary Examination of the Franco-Japanese Case[1]

Patrick A. Messerlin

INTRODUCTION

For many years, in order to induce foreign firms to set up production in Europe, national administrations seem to have been making use of European measures for protection against the import of certain goods. Import restrictions, especially in their most common form of non-tariff barriers, in fact maintain a climate of uncertainty. This in turn induces foreign firms to make direct investments for the local production of goods whose import is threatened. France is no exception to this tendency, which is also particularly marked in Belgium, Ireland, Italy and Spain.

The existence of very strict regulations in France concerning foreign investment authorisation and an elaborate system of capital market controls has, however, encouraged the French authorities to go further than simply apply a classical 'tariff-factory' policy, in order words, a policy of relocation of world productive activities through the manipulation of commercial policy. The French authorities in fact soon sought to impose on the foreign investors precise objectives for employment and above all for exports, this second objective being in order to permit the re-establishment of equilibrium in trade in the product concerned, a criterion regarded as essential by the French authorities. Since 1983, this policy has gone a step further. The aim now is to favour foreign direct investment (FDI) involving cooperation with French firms, and so openly acts in favour of technological transfers.

The most complete example of this policy is undoubtedly provided by relations between France and Japan, especially concerning electronic products. Restrictions on imports from Japan of these products, considered by the French authorities as particularly 'sensitive', have been used most energetically. The decision to make Poitiers the unique customs office for the clearance of video-recorders has become a notorious example of non-tariff barriers. Moreover, the 'technical co-operation' agreements between Japanese and French firms are openly at the heart of the negotiations taking place in the Franco-Japanese Committee for Technical Cooperation.

The aim of this paper is to attempt a preliminary estimate of the economic effects of this complex tariff-factory policy, based on this Franco-Japanese example. The first stage will be to establish what action has been undertaken in the field of French commercial policy as applied to imports from Japan. We shall then try to bring out the consequences in terms of FDI and technical transfers, before attempting to arrive at an estimate of the effects on the French economy in terms of resource allocation and welfare.

FRENCH COMMERCIAL POLICY TOWARDS IMPORTS FROM JAPAN

Trade with Japan accounts for only a small part of total French foreign trade: imports from and exports to Japan are only 13 per cent and 7 per cent respectively of French import and export trade with industrialised countries, excluding the EEC. But imports from Japan have two fundamental characteristics: heavy concentration on a relatively small number of products; and the 'strategic' character, in the eyes of the French government, of the products concerned. The 17 sectors where Japanese penetration is greatest are shown in Table 11.1. Imports in these sectors, on their own, accounted for 30.8 per cent of total French imports from Japan in 1974, 63.3 per cent in 1981 and 58.2 per cent in 1984. One might in fact usefully ask whether this continuing concentration of imports in these sectors, despite the protectionist measures applied, is not an indication of the extent and the deep-seated nature of Japanese comparative advantage in the production of these goods. Examination of the industry breakdown of these products, using the most detailed French industrial nomenclature (NAP 600), shows that 12 of the 17 industries are in the electronics and electronic application fields. Imports of tape/video-recorders (NAP 2922) in 1984 were the largest item, at 3.1 billion francs, followed by office equipment (NAP 2702), data-processing equipment (NAP 2701) and finally tubes and semiconductors (1.9, 1.5 and 1.1 billion francs, respectively).

Table 11.2 lists the principal measures aimed at limiting the import of these products, which come mainly from Japan. The range of measures is very wide. First, there are the customs duties, much heavier than is often thought. Average tariffs range between 4 and 7 per cent, but in practice they are biased towards the lower end of the Common Customs Tariff (CCT) range, since the higher duties probably apply to those imports experiencing a greater volume reduction than those on which duties are lower. The column showing the level of duties in the CCT also shows an average of around 14 per cent for the maximum rates, three times the minimum rate average. This spread, as is well known, is an indirect but fundamental measure of the dissuasive effect being sought.

Then come the quota restrictions in the strict sense of the term, that is to say, managed by the French authorities and giving rise to 'preference for industrialists'; in other words, the allocation of the authorised quotas with priority to domestic producers (Thomson is the latest major one) and, after them, to (non-producing) importers and to foreign producers. Those who carry out production in France are thus assured of being able to import on better terms than foreign producers who have no location in France, as shown by the examples of Akai, Kenwood, Pioneer and Sony. These quotas are particularly prominent in electronics and cars. There are also quotas for particular countries (or groups of countries including Japan), in volume terms, for radios, television sets and video units, with the possibility of setting sub-quotas aimed at improved protection for products thought to be 'threatened'. Of much the same nature as these quotas, but applied only since 1982, are the voluntary export restraint agreements concerning video-recorders which were concluded between the Japanese Ministry of International Trade and Industry and the EEC Commission.

In the case of electronic products, the range of available commercial policy measures also includes the fixing of minimum import prices. This introduces a systematic underestimation of the impact of the customs duties, since these are calculated on the minimum prices and not on the actual prices which the Japanese exporters might be able to offer. Finally, there is the whole arsenal of non-tariff barriers, including, it should be noted, the role played by official norms in electronics and in cars, and, in general, by authorisations from the 'responsible authorities'. For example, since the end of 1985, telephones have to be officially authorised by the

Table 11.1: French imports from Japan and the NICs, 1973–84

NAP 600	Industries	Share of Japan (%)							Share of NICs[a] (%)							Share of Japan & NICs (%)						
		1973	1975	1977	1979	1982	1983	1984	1973	1975	1977	1979	1982	1983	1984	1973	1975	1977	1979	1982	1983	1984
2701	Computing equipment	0.3	1.0	0.8	1.9	2.8	4.2	6.1	0.3	0.3	0.5	0.7	0.6	0.9	2.2	0.6	1.3	1.3	2.6	3.4	5.1	8.3
2702	Office machinery	14.4	14.6	19.7	21.4	25.5	30.1	32.6	1.6	4.1	5.2	4.0	3.6	3.5	3.1	16.0	18.7	24.9	25.4	29.1	33.6	35.7
2911	Telegraphs-phones	4.3	7.1	5.5	2.2	10.9	10.9	10.9	0.1	0.4	0.3	8.1	4.2	4.0	8.0	4.4	7.5	5.8	10.3	15.1	14.9	18.9
2912	Electronics-medical	1.8	2.1	2.6	3.8	5.2	6.2	6.1	0.0	0.0	0.0	0.0	0.0	0.0	0.1	1.8	2.1	2.6	3.8	5.2	6.2	6.2
2914	Electronics-professional	8.3	6.2	8.0	16.3	23.8	14.9	14.8	1.2	1.4	3.3	3.4	2.0	2.5	3.3	9.5	7.6	11.3	19.7	25.8	17.4	18.1
2915	Components	6.7	6.1	5.8	6.7	10.2	9.5	12.7	0.7	0.8	2.1	3.4	3.0	3.2	2.7	7.4	6.9	7.9	10.1	13.2	12.7	15.4
2916	Semiconductors, tubes	0.6	1.6	7.4	9.2	11.3	11.9	12.7	2.2	3.3	3.9	5.3	8.3	9.2	9.7	2.8	4.9	11.3	14.5	19.6	21.1	22.4
2921	Radio and TV sets	12.3	17.8	17.6	17.3	13.5	11.9	11.3	9.8	10.2	12.0	18.4	28.5	30.2	24.8	22.1	28.0	29.6	35.7	42.0	42.1	36.1
2922	Tape & video recorders	8.1	12.9	27.7	43.9	58.8	54.8	52.1	4.7	5.1	4.9	5.8	5.3	4.2	4.7	12.8	18.0	32.6	49.7	64.1	59.0	56.8
3401	Watches	5.4	7.9	13.2	12.9	20.1	23.2	20.4	2.8	2.9	10.6	14.5	13.8	14.0	14.3	8.2	10.8	23.8	27.4	33.9	37.2	34.7
3404	Optical	16.6	21.1	28.9	33.9	31.1	26.1	24.2	2.3	1.8	1.6	1.2	1.2	1.2	1.3	18.9	22.9	30.5	35.1	32.3	27.3	25.5
3405	Picture and movie cameras	26.5	30.8	36.1	34.3	37.2	38.0	34.2	2.7	3.3	3.7	4.0	4.6	4.5	4.4	29.2	34.1	39.8	38.3	41.8	42.5	38.6
3111	Automobiles	2.5	4.6	7.3	6.6	6.0	5.4	6.1	0.0	0.0	0.0	0.0	0.0	0.0	0.0	2.5	4.6	7.3	6.6	6.0	5.4	6.1
3116	Motocycles	43.7	46.9	57.9	62.6	62.2	63.9	60.5	0.0	0.0	0.0	0.2	1.7	0.6	0.6	43.7	46.9	57.9	62.8	63.9	64.5	61.1
1809	Camera and movie products	4.9	5.1	6.7	7.1	9.0	10.6	11.9	0.0	0.0	0.0	0.0	0.0	0.0	0.0	4.9	5.1	6.7	7.1	9.0	10.6	11.9
5405	Musical instruments	19.9	17.3	20.1	15.1	26.1	31.3	17.1	2.2	2.5	4.3	4.6	7.2	7.5	4.4	22.1	19.8	24.4	19.7	33.3	38.8	21.5
5406	Desk equipment	17.4	19.1	19.1	19.4	18.9	18.7	18.6	3.9	7.1	7.5	8.3	8.8	8.6	8.0	21.3	26.2	26.6	27.7	27.7	27.3	26.6

Note: a. NICs = South Korea, Hong Kong, Singapore and Taiwan.
Source: Foreign Trade Department, author's computations.

Table 11.2: French restrictive measures against Japanese imports[a]

NAP 600	Industries	Average customs duty 1974	1981	1984	C.C.T. Mini	Maxi	Declared sensitive	Main non-tariff barriers[b]	Use of article 115[c]
2701	Computing equipment	5.1	4.1	2.7	5.4	7.4	X		
2702	Office machinery	7.5	5.9	4.9	5.2	10.1		Norms. Dumping. DV.	
2911	Telegraphs-phones	4.8	4.3	4.1	4.8	8.1		Agreement DGT. DV.	
2912	Electronics-medical	4.5	4.4	4.1	0.0	12.0		LD. DV & DNV.	
2914	Electronics-professional	4.1	4.2	3.5	5.4	8.7			Japan (6)[d]
2915	Components	4.5	6.3	3.2	5.1	8.7			
2916	Semiconductors, tubes	7.2	6.7	5.6	5.6	15.0	X	LS.	
2921	Radio and TV sets	12.2	7.6	6.3	5.0	14.0	X	Quotas. Min. Price	Japan(16) Korea(12)
2922	Tape & video recorders	5.1	5.8	4.5	5.0	16.0	X	Quotas. Min. Price	Hong Kong(8) Taiwan(12)
3401	Watches	3.8	1.4	1.3	4.8	10.1	X	Quotas	Japan(1) Hong Kong(2) Singapore(1)[e]
3404	Optical	8.3	8.6	6.1	5.5	12.0		LD. DV. Norms	
3405	Picture and movie cameras	8.9	5.4	4.2	8.1	8.8		LD. DV.	
3111	Automobiles	4.3	4.7	4.1	6.7	22.0		Quotas. Norms.	
3116	Motorcycles	9.8	9.1	9.1	6.7	22.0	X	LS. DV & DNV.	
1809	Camera and movie products	5.6	7.6	5.1	8.5	9.1		Min. Price	
5405	Musical instruments	8.2	6.7	6.1	5.4	6.6		Quotas	
5406	Desk equipment	8.2	5.4	4.2	5.7	7.9			

Notes: a. Computed for OECD countries, with EEC countries excluded.
b. DV = controlled declaration; DNV = non-controlled; LS-LD = control or discretionary licences.
c. Applications since 1981.
d. 9028 in Nimexe, the EEC's six-digit trade classification.
e. One case for each of China, Macao, Pakistan and Philippines.
Sources: Foreign Trade Department, customs tariffs, official bulletins, author's computations.

Direction Générale des Télécommunications (henceforth DGT) before they can be imported or sold. One last source of protection is the application of anti-dumping duties, as practised in Europe. These can sometimes be substantial, as shown by the case of the Japanese producers of electronic typewriters, who found themselves in December 1984 having to pay countervailing duties of 20 to 34 per cent.

The shares of Japanese products in French imports cannot really be appreciated without some reference to the newly industrialising countries (NICs) in the Far East (South Korea, Hong Kong, Singapore and Taiwan). These countries have developed export strategies enabling them from the end of the 1970s to be competitive with Japan in a certain number of industries considered here. Some French firms have participated in the expansion taking place in these countries, for example Thomson in Singapore in the field of mass market electronic goods. To a certain extent, the fact that French firms have set up in the NICs has prevented the application of as strict a policy of import restrictions as in the case of Japan. This is why it was possible to see a substitution of the NICs for Japan, starting at the end of the 1970s, in the case of television sets, where strict quota restrictions were applied towards Japan. This substitution of new exporters for Japan, combined with differences in the treatment of imports from third countries by EEC members, sets limits on the effectiveness of the restrictive measures applied. Nevertheless, as shown in Table 11.2, Article 115 of the Treaty of Rome, which allows for the banning of imports 'in free circulation', i.e., coming from a country subject to quotas but transiting through another Community country, has been systematically used by France since 1981. Two comments deserve to be made at this point. First, Japan is an undoubted 'beneficiary' of applications under Article 115, accounting for more than 40 per cent; second, it is the country for which the applications are the most frequently rejected by the Commission (about 50 per cent of cases), suggesting that the Article is being used as a means of pressure and 'commitment'.

FINANCIAL AND TECHNICAL TRANSFERS FROM JAPAN TO FRANCE

Table 11.3 recapitulates all the major Japanese investment operations in France from 1968 to 1984. Most are in fact quite recent: out of 26 operations enumerated, 20 have taken place since

Table 11.3: Japanese Investments in France, 1968–84

NAP 600		French subsidiaries	Main sectors	Date	% Equity	French partners	Number of employees	Location
132	Suntory	Chat. Lagrange	Wines	1983	100.00		50	Medoc
1310	Toyo Alumin.	Alcan Toyo Eur	Aluminium	1982	50.00	AA France	50	Accous
1721	Ajinomoto	Eurolysine	Lysine	1976	50.00	Lafarge-Coppée	260	Amiens
1727	Dai Nippon	Nordic SA	Plastics	1978	49.00	Normetex		Nantes
1804	Alfa Techno		Glues	1984	50.00	Rousselot Prod.	50	Ibos
1807	Dai Nippon	Georget	Inks	1980	35.00	Ripolin	250	Nantes
1811	Sagami Gum	Radiatex	Perfumeries	1983	100.00	Repurchase option	60	Vichy
2702	Epson	Epson	Printing	1983	100.00		50	
2702	Canon	Canon Bretagne	Copiers	1983	100.00		180	Rennes
2702	Canon	Canon Bretagne	Typewriters	1984	100.00		140	Liffre
2921	Unitec	Dicorop		1982	50.00	Dicorop	150	Cannes
2921	Clarion	Clarion	Auto-radio sets	1983	51.00	M. Bessis	200	Poapey
2922	Sony	Sony-France	Magnetic tapes	1980	100.00		400	Bayonne
2022	Akai	Akai-France	Tuners VTRs	1982	70.00	Akai France: 30%	200	Honfleur
2922	Sony	Sony-France	Video-K7	1982	100.00		450	Dax
2922	Pioneer	Pioneer France	Baffles	1983	70.00	M. Setton	80	Bordeaux
2922	Trio Kenwood		Tuners Hifi	1984	50.00	DeltaDore & SDR	75	Rennes
2922	Akai	Akai France	Magnetoscopes	1984	100.00		330	Honfleur
3113	Stanley Elect.		Tables LCD	1984	31.00	Renault: 45% X: 24%	200	LeBourget du Lac
4709	Yoshida KK	YKK	Zippers	1968	100.00		300	Lille
4804	Astre Deco		Pannels	1984	100.00		100	Carcassonne
5201	Sumitomo	Dunlop France	Tyres	1984		Repurchase option	3500	Several
5402	Daiwa Seiko		Carbon	1984	100.00		145	St Etienne Rouvray
5406	Pentel	Pentel France	Pens	1970	100.00		50	Bry
9999	Toray	Soficar	Carbon fibres	1982	35.00	Elf & Pechiney: 65%	180	Pau
9999	JMA Japan	JMA Consultant	Consultants	1984	100.00		30	Paris
							7480	

Sources: DATAR, L'Usine Nouvelle (13–12–84), La Tribune (08–03–85).

1982 (5 in 1982, 6 in 1983 and 9 in 1984). Grouping these direct investments by industry immediately shows that most of them involved electronic and related industries, with 12 of the 26 in this category if one includes the manufacture of liquid crystal displays by Stanley for the car industry. Out of the 3,980 job creations announced (excluding the take-over of Dunlop by Sumitomo), the electronic industry accounted for nearly two-thirds or 62 per cent, including 39 per cent in the production of tape/video-recorders (NAP 2922), and 9 per cent for radio and television sets (NAP 2921) and for office equipment (NAP 2702). Table 11.3 contains two further items of information. First, there have been no take-overs in electronics. Second, Japanese investors have had no hesitation in setting up factories away from the Paris area, thus confirming a well-known tendency in the case of foreign investment in France.

Thus Table 11.3 underlines the impact of the French authorities' tariff-factory policy. Japanese investments have been most numerous in cases where (and in the period since) the most severe protectionist measures have been applied.

In order to establish a clearer picture of the totality of the effects of these direct investments on the French economy, it is essential to compare the two industries principally concerned with other industrial groupings taken as controls. For this purpose it has been necessary to use an industrial nomenclature which is less detailed than the one used so far, and to consider just two electronic industrial groupings: NAP 27, which comprises NAP 2701 and 2702; and NAP 29, comprising NAP 2911 to 2922. The two control groups are those with a high proportion of foreign firms and those with exportable products.

The industrial characteristics selected for Table 11.4 fall into two groups. The first concern intensity of factor use and are, therefore, based on the pure classical theory of international trade. For example, value added per worker reflects the need of the industry in question for 'capital' in all forms — physical, human and natural resources. This summary indicator of factor intensity is accompanied by others which are more refined but also more volatile. The general impression derived from the table is that French export industries are more capital-intensive than the economy as a whole, although the indicator of R&D expenditure shows how necessary it is to distinguish between physical and human capital. The second set of indicators has been introduced to take account of the variables more commonly used in recent international trade theories and associated with imperfect competition, economies of scale and

Table 11.4: Industrial environment of Japanese investments, 1980

Indicators	Export industries	Industries with a high degree of foreign investment	Office equipment industries (NAP100:27)	Electronic equipment industries (NAP100:29)
Factor intensity				
Value added/per worker	0.78	1.24	1.32	0.72
Capital stock/per worker	0.88	1.06	1.99	0.40
Employment cost/per worker	0.98	1.11	1.67	1.00
RD expenses/per worker	1.31	1.72	4.24	5.89
Labour intensity[a]	1.02	0.96	0.97	1.06
Industrial structure				
Concentration[b]	0.80	1.01	2.57	0.83
Vertical integration[c]	0.94	1.03	1.28	1.10
Scale economies[d]	0.96	1.07	1.18	0.63
Differentiation[e]	1.10	0.83	0.62	1.25
Share of foreign firms	1.24	1.43	1.40	1.51
Share of public firms	0.96	1.26	0.43	2.27
International trade				
Degree of openness	0.90	1.07	1.59	0.92
Penetration rate	0.62	1.25	2.23	1.40

Notes: all indicators are calculated on the base 1 for the whole economy.
a. Employment cost/value added.
b. C_4 revised by index for four firms.
c. Turnover/value added.
d. Hufbauer's coefficient.
e. Number of NAP 600 categories.
Source: Messerlin (1984).

product differentiation. From the point of view of this second set of variables, the exportable-goods industries stand out by their relatively low concentration ratio and the high proportion of foreign firms.

Table 11.4 enables the two industries to be ranked in relation to the two control groups and, in this way, provides an answer to two fundamental questions. First, it makes it possible, given the extraordinary concentration of Japanese investments in certain industries, to examine how typical or atypical they are of other FDI in France. There is a significant positive correlation between the indicators for NAP 27 and 29 and those of all industries with a high proportion of foreign firms. This result tends to refute the idea that Japanese investments in France are atypical and suggests that, although French commercial policy may have put pressure on the Japanese investors, the nature of the investments has been fully in keeping with the fundamental economic forces in operation. Second, by taking as control group the exportable-goods industries, we can examine whether the industries where there is Japanese penetration are among those where there is French comparative advantage. What was said earlier about French commercial policy would lead one to expect the absence of any significant correlation. This turns out to be true for NAP 27, but in the case of NAP 29 there is in fact a significant positive correlation. However, this result could stem from the fact that NAP 29 includes both professional electronic goods, in which France is often regarded as being competitive, and household electronic goods. In addition, this NAP is often the subject of export credit financing, which would introduce bias into the indicators of comparative advantage.

A TENTATIVE ESTIMATE OF THE EFFECTS OF THE POLICY

What effect have the limitations on imports had on French production, excluding Japanese establishments? This effect can be shown by two variables: first, changes in the volume of imports and, second, changes in the prices of domestic products. Under the voluntary restraint agreements, the Japanese undertook to limit their exports of video-recorders to 450,000 units in 1984, whereas the trend in French imports would have allowed them to hope to export a larger amount, at least 550,000. Actual imports in 1984 were 490,000 units, 60,000 less than they could have been.

The reaction of the prices of the products manufactured in France to the rise in the prices of imported Japanese video-recorders is the result of two opposing forces: on the one hand, the rise in production costs to meet increased production needs; on the other, the effects of competition between European producers. It is not obvious that these two forces operated very strongly, especially since consumers were worried by having to choose between different standards and since the two main European firms, Philips and Thomson, took a long time to decide on their strategies and launched into a take-over race of dubious value. In addition, examination of the price indices does not show any clear price movement following the import limitations, given that INSEE, France's National Institute of Statistics, has no separate price index for video-recorders, but only one comprising all mass-market equipment other than radio and television sets. This lack of solid empirical evidence leads us, therefore, to take a prudent attitude for the present and to assume that the rise in import prices did not lead to a rise in domestic prices in 1983–4, but restricted the fall which could otherwise have been expected. This provisional hypothesis probably leads to an underestimation of the increase in domestic production resulting from the import restrictions.

The increase in domestic production resulting exclusively from the fall in imports can then be estimated by the following equation:

$$dV/dM = (V/M) \cdot (Ev/Em)$$

where Ev is the elasticity of demand for domestic products and Em that for imports (both in relation to import prices), and where V represents domestic sales and M imports. Since Ev is generally estimated at 0.60 and Em at about 1.15, this means that French production increased by about 1,000 video-recorders as a result of the non-import of 60,000 Japanese video-recorders. This is a somewhat disappointing result, but apparently in line with reality.

These gains in domestic sales were not cost-free. Two main types of cost can be considered in order to estimate the net gains of the tariff-factory policy. In the first place, economic theory leaves little doubt as to the consequences of a protectionist policy. While such a policy can increase the revenue of the government and temporarily stabilise the situation of the marginal domestic firms by sheltering them from international competition, it obtains these positive effects only to the detriment of the situation of the consumers. In practice, the economy incurs two welfare losses. The first stems from the fact that consumers cannot buy all the imports they would like; the second, because the domestic economy is induced into the wrong kind of specialisation, in the sense of producing goods which it is unable to manufacture as efficiently as its foreign competitors. This picture only applies in situations where there is competition. But it is difficult to see France enjoying the monopoly power for electronics which would enable it to compensate these net losses by gains on the terms of trade. In addition, although dominated in the case of some products by very large firms, the electronic industries are apparently subjected to strong competitive pressures linked to the very rapid evolution of certain products, such as video-recorders during the past ten years or, most likely, television receivers during the next ten.

In order to consider the second type of cost, one must ask how far this traditional analysis is still applicable when flows of imports subject to artificial limitations are replaced by movements of capital in physical, financial or technical form. Until recently, economists generally felt that the beneficial effects of these movements offset the negative effects of the protectionist barriers. This opinion was based on the theorem of Mundell (1957) proving that imports of factors allow an autarkic economy to re-establish the relative prices of goods and factors, and especially the level of welfare that it would have experienced under free trade with no movement of factors between countries. It should be stressed here that Mundell's theorem is an indifference theorem, at least from the point of view of the country concerned: the country returns to the welfare situation under free trade, neither better nor worse off. The imports of foreign factors of production have given the country the factor endowment corresponding to its desire for specialisation. In other words, this approach pays no attention to the authorities or to the value which they, therefore, attach to one type of specialisation rather than another.

More recently, however, this opinion, widely accepted by economists, has been called into question in articles by Jones (1984) and Neary and Ruane (1985). Each of these contributions shows, on the most general assumptions, the negative effect exercised by capital movements triggered off by protectionist measures. The reason is simple: this additional capital would only take the economy further down the road of the 'wrong' specialisation, by drawing more domestic resources into the production of goods which the economy would import if it were not preventing from doing so by import restrictions. This is, therefore, the second type of cost which should be introduced in order to arrive at the net gain from a tariff-factory policy.

The measurement of the cost to the consumer of the adoption of protectionist measures merits a study of its own and will not be attempted here. The most complete treatment of the gains to the French economy from Japanese investments demands answers to the three following questions. First, what prices should be used to calculate the production by Japanese establishments in France? Obviously, these should be international rather than current French prices, since this production has to be regarded as substituting for imports. To estimate the additional sales by factories set up by Japanese firms, it would therefore be desirable to weight them by Japanese export prices, for example. Second, what type of skills are Japanese factories generating in France? The fact that highly-skilled jobs are being created is not a sufficient answer in itself. The more important question concerns the degree of specificity of the skills resulting from the Japanese establishments: the abolition of import restrictions will be easier if the job skills are unspecific, since this would reduce the costs of adjusting to this desirable step. Finally, what effects has the setting up of Japanese firms had on the efficiency of French firms' production methods? The technical innovations at the product and manufacturing levels have to be included in any calculation of the gains from this tariff-factory policy, to set against the losses mentioned earlier. Once again, it is important to know whether this last type of gain has been obtained at international cost levels or higher ones. It is to the measurement of these three points that any complete study of the effects of Japanese investments in France must devote itself.

NOTE

1. March 1986.

12

Technology Transfers in the Automotive Equipment Industry: the French Case

Daniel Soulié

The automotive equipment industry has several specific features, three of which are closely linked to its present evolution. The first is that automotive equipment is produced by two diferent types of firms, specialised equipment makers and car makers. Second, there is a high rate of technical change within the industry, both for products and processes. Third, there is standardisation on a world scale of some segments of the car market. This phenomenon is known as 'market globalisation'. The globalisation of the car market implies that of related automotive equipment. The industry's evolution has led to changes both in the nature of the links between car and equipment makers and of the management of their business relationships.

It appears that French firms were unwilling or unable to cope efficiently with these major changes. This has resulted in unfavourable consequences: technological advantages are disappearing, the firms' international strategic positions are weakening and there is a comparative disadvantage in management costs. This pessimistic statement is corroborated by a study of the situation of the French automotive equipment industry in the international technology transfer process. This is true of 'hard' technology transfers, concerning products and processes, as well as of 'soft' technology transfers, dealing with management and strategic behaviour.

Though most studies on technology transfers are related only to the hard aspect of the problem, the interdependency between equipment and car makers is so tight that it is impossible to dissociate hard from soft technological transfers. Therefore, this paper deals sequentially with both aspects of the technology transfer process.

HARD TECHNOLOGY TRANSFERS: PRODUCTS AND PROCESSES

From a very general point of view the weakening of the French technological position in the field of automotive equipment originates in the inadequacy of R&D activity: either spending is too low, or it proved impossible to make it profitable.

Except for a very few firms, French automotive equipment makers are not large enough to reach the critical threshold in the field of R&D. Moreover, the relationship with their buyers, the car makers, which is too often of a conflicting nature, has had two results. First, profitability is low. Second, it appears difficult to develop a vertical joint activity in the field of R&D, though it would most likely prove to be efficient. Therefore, even for large firms, R&D budgets are quantitatively and qualitatively inadequate. Their amount is not high enough to guarantee the survival of firms confronted with harsh international competition. In addition, resources are often inefficiently allocated.

Automobile producers cannot expect to undertake and finance by themselves all the R&D needed to develop both car and equipment industries. Moreover, at the present time, their profitability is low and their financial situation troubled. There is also a specific economic problem with equipment. Naturally enough, car makers are not eager to purchase equipment, especially sophisticated items, from an integrated car maker. Therefore, it is very difficult for an automobile producer to make profits from R&D expenditures in equipment. This is most likely the reason why Renault had to withdraw from Renix, a joint venture in automotive electronics it developed with Bendix.

The unfavourable evolution of the balance on hard technology transfers reflects the declining performance of French industry in this field. Technology transfers are difficult to measure directly and precisely. Therefore, we use an indirect approach, measuring three main indicators: the degree of advanced technology in French international trade, the origin of patent applications filed in France, and the payments made for technology transfers.

International trade and technological performance

Our basic idea is that the high technology content of international trade reflects the situation of French industry compared to that of its main competitors. Roughly, the situation of the French automotive equipment industry is good if France exports high-technology products and imports standardised products. Conversely, if the balance of high-technology international trade is adverse, France should be considered as lagging behind other industrialised countries.

Putting this principle into practice is more difficult than expected. One problem is that the concept of a high-technology content in exports and imports is mainly qualitative. Furthermore, the analysis is meaningless if made in terms of absolute values. It must be developed in terms of evolution of the balance. This latter approach implies a restrictive assumption: for the whole period under study, products must be assumed to retain the same (relative) amount of high-technology content. For these reasons, our sample consists only of products for which the existence of a high-technology content is unquestionable.

Due to the qualitative dimension of the problem, we had to rely on expert advice. As a first step, we asked the advice of two experts regarding the French customs classification for automotive equipment. The question was to enter each product in one of the three following categories: high-, medium- or low-technology content. As a precaution, we took the three following measures: there were no figures of international trade on the list of products, experts were interviewed independently, and they were not informed of the existence of a second interviewee. Comparing the independent answers, it was possible to draw a list of six equipment categories considered by both experts as having a high-technology content. As a second step, we studied the international trade of these six groups of equipment. The main results appear in Table 12.1.

It appears that there is great variety in the absolute level and the evolution of foreign trade for the different equipment groups. There does not seem to be a general rule of evolution in this type of international trade. There is, however, a clear deterioration of the situation of the French automotive equipment industry over the period. The overall balance for the six categories is still positive in 1983, but, for five of the six the ratio of exports to imports is lower in 1983 than it was in 1973.

Categories within the customs classification are not always very significant from an economic point of view since some categories cover different types of equipment. Moreover, in some cases the data collection may not be reliable. Nevertheless, the evolution of high-technology external trade is worrying. Although the overall

Table 12.1: International trade in high-technology products

Millions of constant (1970) francs	Exports A		Imports B		A/B		Average yearly rate of growth	
	1973	1983	1973	1983	1973	1983	Exp.	Imp.
Safety and electric locks for automobile cylinder blocks	14.39	18.00	8.95	12.96	1.61	1.39	2.3%	3.8%
Cylinder blocks, cylinders, sleeves, crank cases and cylinder heads for internal combustion engines	108.39	38.90	87.76	76.96	1.24	0.51	−9.7%	−1.3%
Push-rods, piston rings for internal combustion engines	15.57	20.42	31.69	42.98	0.49	0.48	2.7%	3.1%
Carburettors, injectors and injector holders for internal combustion engines	53.10	196.25	80.04	101.68	0.66	1.93	14.0%	2.4%
Crankshafts and camshafts	25.11	63.70[a]	15.25	51.21[a]	1.65	1.24[a]	10.9%[b]	14.4%[b]
Ignition and starting equipment for internal combustion engines	112.14	235.98	28.85	105.80	3.89	2.23	7.7%	13.9%

Notes: a. 1982. b. Period 1973–82.
Source: French customs statistics.

external trade in automotive equipment is still satisfactory, the adverse evolution of high-technology segments is an advance indicator of the weakening of the industry within the context of international competition. It will probably be followed by a deterioration of the overall trade balance.

The same pessimistic results are given by the study of patent applications and technology transfers.

Innovations and international technology transfers

Studying the juridical and financial aspects of property rights in technology is another way of gaining a view of the international position of the French automotive equipment industry. Data on two indicators were available: patent applications and technology transfers.

Patent applications

Most scholars agree that the number of patent applications filed by a firm is an imperfect measure of its innovative behaviour. The filing of patent applications requires time and money, administrative procedures are complex, and juridical protection precludes secrecy. Therefore, the propensity to patent inventions and discoveries is less than one, and probably steadily decreasing.

In the absence of more relevant data, consideration of the number of patent applications leads to a rough estimate of what happens in the automotive equipment industry.

Table 12.2 features data on patent applications for the industry as a whole and for major categories of equipment, regardless of the nationality of applying firms. There is a steady decrease in the number of applications, but this is probably due to institutional rather than economic reasons. Only applications filed through the national (French) channel are taken into account. Foreign firms tend to make increasing use of the European channel, but related data were not available with a breakdown by products. Taking into account applications through the European channel leads to a much smaller decrease.

When considering a breakdown of applications by nationality and field of activity of firms, instead of products, it appears obvious that French industry lags behind international competitors. In Tables 12.3 and 12.4 the data refer only to major applicants with over 10 applications per year. Foreign firms filed twice as many applications

Table 12.2: Patent applications filed in France

	Class B60	French channel only 1980	1981	1982	1983
B	Wheels, axles	89	101	70	59
G	Suspensions	73	67	59	58
H	Heating	72	45	47	47
J	Windows, doors	121	106	95	113
K	Front and rear wheel drive equipment	278	272	245	172
L	Electrical equipment	46	34	26	22
N	Miscellaneous fittings	111	95	90	78
Q	Lights	99	72	68	41
R	Others	373	279	214	197
T	Brakes	225	204	105	133
	Sub total	1,487	1,275	1,064	920
	Total, Class B60	1,882	1,638	1,240	1,151

Source: Institut National de la Propriété Industrielle.

Table 12.3: Major French applicants for patents[a]

Equipment makers	1981	1982	1983
Valéo	61	94	85
Equipement Marchal	23	19	27
DBA	34	42	98
Ducellier et Cie	32	23	18
Accumulateurs fixes et de traction	—	—	15
Cibié projecteurs	—	17	12
Ferodo SA francaise[b]	32	—	—
Usinor	19	—	—
Jaeger	—	18	—
Neiman	—	17	—
Glaenzer Spicer	—	10	—
Total	201	240	255
Auto makers			
Peugeot automobiles	34	40	49
Renault	117	84	113
	34	24	26
R.V.I.	—	—	23
Total	185	148	211
TOTAL	386	388	466

Notes: a. Lack of figures does not imply that no application was made, but that the number of applications was less than 10.
b. Part of Valéo since 1982.
— Data not available.
Source: Institut National de la Propriété Industrielle.

Table 12.4: Major foreign applicants for patents[a,b]

Equipment makers	1977	1978	1979	1980	1981	1982	1983
Bosch (FRG)	324	299	253	—	194	239 (88)	261(111)
Lucas (GB)	149	98	109	96	98	130 (30)	91 (37)
Bendix (US)	56	98	107	27	—	94	81 (76)
Motorola (US)	46	36	—	—	—	—	—
Automative Products PLC (GB)	—	—	—	—	(33)	—	—
Total	575	531	469	123	325	463	433
Auto makers							
Nissan Motors (J)	63	56	206	291	128	241(212)	156(141)
Honda (J)	—	43	66	47	51	90	74
Daimler Benz (FRG)	90	101	88	111	88	61	64
General Motors (US)	—	—	47	51	—	29	58 (47)
Fiat (It)	—	44	—	—	—	28	39
BMW (FRG)	—	—	—	—	—	43 (37)	36 (36)
Porsche AG (FRG)	—	—	—	—	—	25	35
Ford Motor Cy Ltd (GB)	—	—	—	27	(45)	—	—
Ford France (F)	—	—	—	—	10	—	—
Ford Werke AG (FRG)	—	—	—	—	—	— (47)	—
Total	153	244	407	527	322	564	462
TOTAL	728	775	876	650	647	1027	895

Notes: a. Lack of figures does not imply that no application was made, but that the number of applications was less than 10.
b. Numbers in brackets are related to applications through the European channel (European Patent Agency).
— Data not available.
Source: Institut National de la Propriété Industrielle.

as French firms between 1981 and 1983, on the average. Though the indicator is rough, this result shows clearly the deterioration of the French automotive equipment industry's technological position.

The same conclusion applies to the study of international payments for technology transfers.

International technology transfers

Unfortunately, there are no data available on international technology transfers of equipment alone. Data on international payments for technology transfer relate to the automobile industry as a whole, including equipment. A rough estimate is that, for most years, payments for equipment technology transfers amount to between 40 and 60 per cent of total payments, both for expenses and receipts. There are three exceptions to that rule; during the years 1974, 1978 and 1981, a French automaker repatriated large amounts of royalties earned through a Swiss subsidiary.

There is a clear trend in Table 12.5 towards a deterioration of the balance of payments for technology transfers. Fluctuations may be wide from year to year, but taking into account equipment only (that is, correcting for the repatriation just noted) the gap between expenses and receipts tends to widen. The evolution of this third indicator confirms the pessimistic conclusions reached in the two first studies. The French automotive equipment industry was not able to follow technical changes and to develop a policy innovative enough to keep abreast of its major foreign competitors.

This survey of the recent past shows unambiguously that, in the field of hard technology, the French automotive equipment industry lags more and more behind that of other industrialised countries. The same phenomenon is to be found in the field of soft technology. French firms appear to have a comparatively low rate of innovation in management. Moreover, they prove unable to adopt the most efficient procedures implemented by foreign firms.

SOFT TECHNOLOGY TRANSFERS

The French automotive equipment industry has met two major obstacles in the field of management technology transfers. The first is the existence of conflicting relationships between auto makers and equipment makers. The second is the too-important role of auto makers' research departments in the design of automobiles and of their parts. Although these two problems may appear to be indepen-

Table 12.5: Technology transfers, French automobile industry, 1973–83 (thousands of current francs)

	1973	1974[a]	1975	1976	1977	1978[a]	1979	1980	1981[a]	1982	1983
Expenses (A)	54,000	43,000	83,000	136,000	184,000	189,000	222,527	216,448	115,650	134,455	202,918
Receipts (B)	65,000	178,000	158,000	108,000	174,000	265,000	148,695	109,252	540,590	128,066	148,432
Balance	11,000	135,000	75,000	−28,000	−10,000	76,000	−73,832	107,196	425.350	−6,389	−54,486
$\frac{B}{A}$ %	120.4	414.0	190.3	79.4	94.6	140.2	66.8	50.5	467.8	95.2	73.1

Note: a. Repatriation from Switzerland by an auto-maker. The positive balance is therefore not related to equipment.
Source: Institut National de la Propriété Industrielle.

dent, actually they are closely linked. Both lead to the fact that equipment makers are dominated by auto makers. Equipment makers lack the freedom which is necessary not only to develop strategic management but even to implement efficient day-to-day management procedures. Equipment makers are not considered by automobile firms as being responsible.

In a context of slow economic growth, this situation results in adverse economic and financial results for equipment makers, and, through them, on the situation of the entire automobile industry. One major result is too many models for each piece of equipment. Automobile firms' research departments want each piece of equipment to be technically fitted to the model of car (or version of a model) it will equip.[1] Moreover, equipment is often designed from an auto maker's point of view, that is, regardless of economic or technical problems raised by its own production. This leads to higher production costs than there would be if equipment were standardised and designed jointly by equipment producer and auto maker.[2] This leads as well to very high inventory costs.

Price-setting policies have an adverse impact on equipment makers. First, the so called 'double price' system distorts the market. Equipment fitted on new cars is subsidised by replacement equipment. Skimming the cream on the replacement market is, therefore, attractive for independent competitors, and cannot be legally prohibited. Second, the decrease in sales leads French auto makers to resort to any means of reducing their production costs. They are especially eager to maintain low supply costs, and often succeed in imposing low prices on equipment makers. The lack of financial resources leads the latter to implement short-term survival policies, instead of developing long-term strategies. As noted above, there are not enough investments in technical R&D. Resources are also too scarce to allow investment in improving management techniques.

The solution does not lie in simply imitating procedures used abroad. Some of them would not be efficient within the French economic and cultural context. It is important not only to try to adapt such procedures to French conditions, but also to be able to innovate in the field of management. A prerequisite is an atmosphere of co-operation between equipment and auto makers. There is, at the present time, a real opportunity for relationships to evolve in that direction, including suggestions about the implementation of a system of 'partnership'.

The adverse evolution of the international technological position of the French automotive equipment industry is both the source and a consequence of poor economic performance. Breaking up that vicious circle won't be an easy task. A major effort is needed, both in the technical and managerial fields. A prerequisite for success is a change in the behaviour of both equipment and auto makers.

NOTES

1. As an example, a single firm had to produce 37 different models of fuel pumps in order to equip the same engine, though four models would have been enough. The same engine has been equipped with 17 different models of carburettors.

2. One French equipment maker providing the same kind of product to French and foreign auto makers estimates its production cost differential, due to standardisation, to be at least 20 per cent in favour of its foreign customer. Of course, this estimate may vary according to the kind of equipment.

13

The Japanese Productivity Advantage in Automobile Production: Can it be Transferred to North America?

Melvyn Fuss and Leonard Waverman

INTRODUCTION

A number of joint ventures (JVs) have been announced recently in the automobile industry. These JVs exist for a number of reasons. One such reason is that shared development and production facilities among smaller producers can lead to shared design costs and scale economies in production. Examples of design and production sharing JVs exist. Renault (which owns 46 per cent of American Motors Corporation (AMC)) and AMC are engaged in JVs in Canada and the USA. Outside the USA, Nissan has JVs with both Alfa Romeo and Volkswagen. Honda and British Leyland have a JV as well. In addition, there are numerous supply arrangements and JVs to produce engines and transmissions among a host of other European firms. In this paper we examine the issue of the JV as a means of transferring technology. That reason was paramount in the US government's approval of a JV between the first and third largest automobile producers in the world. General Motors and Toyota are jointly producing a Toyota-designed car in Fremont, California, with major components imported from Japan. This car, the Nova, will reach an output level of 240,000 cars in 1986 and be sold exclusively by GM. This JV was the subject of detailed scrutiny by the FTC as well as the subject of a recently settled private anti-trust law suit initiated by Chrysler.[1] Chrysler has recently announced its own JV with Mitsubishi to produce a Japanese-designed car in the USA. Ford also will produce 200,000 units per year of a Mazda-designed car in Mexico, likely to be for sale in the USA. Mazda has announced a car production facility in the USA with a 240,000 unit annual volume, some of which will be sold by Ford.[2] Each of these investments involves a Japanese firm designing a US-based assembly and stamping plant with major auto components (engine, transmission, transaxle, etc.) imported from outside the USA.

GM contended that cost-competitive production of small cars in the USA was impossible without the transfer of the Japanese technology to US producers. Furthermore, GM argued that this transfer of technology could not occur either by hiring Japanese managers or by visiting Japanese car plants. Foreign firms have been unsuccessful in attempting to hire Japanese managers or automotive engineers away from Japan due to the cultural ties and lifelong contracts which restrict the mobility of Japanese professionals. While US and European auto firms have toured Japanese auto plants, GM contended that lower production costs could only be obtained through designing and working on a production line with the Japanese.[3]

The JVs between Japanese and American firms are suggested as necessary in order for North American producers to learn how to produce small cars. The need for this 'learning' is shown by the alleged large cost and productivity advantage of Japanese auto producers over their American counterparts. Abernathy, Clark and Kantrow (ACK, 1983) suggest a production cost advantage for the Japanese of some US $2,000 per vehicle — 30 per cent of the US producers' costs (these figures are corroborated in other studies: Abernathy, Harbour and Henn (AHH, 1981); Perry (1982); Federal Trade Commission (1983)).

In this paper we report on a larger study of ours which analyses the evidence which supports the view that Japanese auto producers have a large productivity advantage (Fuss and Waverman, 1985, 1986, 1987). We provide our own estimates based on this literature and examine JVs as a form of technology transfer designed to increase productivity. We find, first, that the Japanese cost advantage has been exaggerated in earlier studies; and, second, that much of the so-called cost advantage is either due to short-run temporary differences or due to factor-cost advantages to the Japanese producers. These results suggest only small productivity gains from transferring Japanese technology to North America via a JV, calling into question the desirability of allowing these JVs, given their impact in raising concentration. Furthermore, other studies have suggested that Japanese technology may not be efficient at North American factor prices diminishing even further the potential cost-reducing effects of JVs.

INTERNATIONAL DIFFERENCES IN PRODUCTIVITY AND WAGES: THE 1950s TO THE 1980s

What advantages of production in one country can be transferred to a second country with a JV? In analysing the potential of JVs to reduce costs we must carefully distinguish between differences in costs and diferences in productivity and also examine the reasons for these differences. Of prime importance is the difference in the price of labour (or the price of other inputs such as capital or materials). Differences in these factor costs are not differences in productivity nor can they be 'transferred' to another country. In an industry such as auto production with high capital-output ratios and long-lived capital, diferences in capacity utilisation between two countries will yield differences in costs; these resulting cost differences are short-run and may not be removed by technology transfers. In addition, cultural or educational factors which affect labour productivity cannot be transferred. What can be transferred is the technological organisation of production, the handling of labour, and management techniques.

It is instructive to remember that the 'Japanese advantage' in the production of automobiles and other manufactured goods is a relatively new phenomenon. In the mid to late 1950s, most Japanese manufacturing and the Japanese motor vehicle industry in particular faced large scale and productivity disadvantages as compared to US manufacturing (Bain, 1966).

In contrast, in the late 1970s a number of studies show *large advantages* to Japanese automobile producers over their North American counterparts. For example, a 1978 study by Toder *et al.* indicates that in 1974 the Japanese automotive industry had 17 per cent lower total production costs than did US automobile producers,[4] while Japanese labour costs were some 42 per cent of those in the US auto industry (implying, therefore, less efficient production in Japan or more costly inputs other than labour). Based on these analyses, Toder *et al.* concluded that the US automotive industry was in severe difficulty in the mid to late 1970s facing a Japanese comparative advantage.

Table 13.1, based on AHH (1981), is a summary of a more recent analysis of comparative automobile costs and productivity in the US and Japan and the sources of cost and productivity differences. The authors concluded that in 1979 Japanese auto makers had a *$1,650* total landed cost *advantage* over US producers (and a $2,100 advantage ignoring transportation costs). This differential was mainly due

Table 13.1: Comparative costs and labour productivity in selected United States and Japanese automobile companies

Productivity/cost category	Ford	GM	Toyo Kogyo	Nissan
Labour productivity[a]				
Employee hours per small car	84	83	53	51
Costs per small car				
Labour	$1,848	$1,826	$ 620	$ 593
Purchased components[b] and materials	3,650	3,405	2,858	2,858
Other manufacturing costs[c]	650	730	350	350
Non-manufacturing costs[d]	350	325	1,100	1,200
Total	$6,498	$6,286	$4,928	$5,001

Based on AHH (1981).
Note: Non-manufacturing costs include the costs of ocean freight (for the Japanese producers), selling, and administrative expenses. Other manufacturing costs include costs of warranty, capital costs, energy costs, and miscellaneous items like insurance.
Sources: a. US: 1979 annual reports on numbers of workers (domestic) x average hours per person. Total hours worked divided by domestic production. Adjusted GM for higher vertical integration (value added to sales, .54) by multiplying total hours by (.4/.54). Estimated hours per car size by using *1974* data (corrected to 1979 costs) on *relative* costs of different sized cars (see Toder *et al.*, 1978). Used product mix strategies to give hours per car.
Japan: Company data X1.15 (even though value added to sales = .40).
b. US: Company reports data adjusted for vertical integration (as above). Corrected for product mix (index 1.00 — small, 1.35 — medium, 1.71 — large).
Japan: Martin L. Anderson 'Strategic Organization of the Japanese Automotive Groups', M.I.T. (unpub. 1981), data for Nissan for 1978, valued at exchange rate of 220 yen/dollar.
c. Contains energy, depreciation, warranty costs and miscellaneous. Energy costs in the US assumed to be $150 per car. Used Anderson data for Nissan and a report by J.E. Harbour ('Comparison and Analysis of Manufacturing Productivity', Final consultant report, Harbour and Associates, Dearborn Heights, Michigan, 1980).
d. Contains shipping, marketing and distribution. Sources are 'Most Wall Street analysts and industry executives cite a figure of $400 for the costs added to a Japanese car by ocean freight and US tariffs' (AHH, p. 139).

to labour cost differences since the authors found that the Japanese had a *disadvantage* for other components. Utilising a plant-by-plant comparison, they estimated that in 1979 the Japanese automotive industry required 80.3 man hours to produce a small automobile while the US industry required some 144 hours to produce a similar vehicle. If these numbers are correct (and we disagree), and if the higher Japanese labour productivity is due to a transferable technological feature of production (and not the result of work ethic), then a JV which transferred this technology could raise productivity in other countries.

A 1983 study by Perry updates the AHH data to 1980 and calculates a 31.5 per cent one-year increase in output per worker in Japan relative to that in the USA. This estimate would appear to be erroneous, as would any comparison of labour productivity between the US and Japan in 1980. In that year, the Japanese automotive industry was utilising 102 per cent of capacity while the US industry was operating at only 58 per cent of capacity. This substantial difference in capacity utilisation suggests that measured labour productivity in 1980 represents short-tun disequilibria as well as any fundamental differences in efficiency between the two countries' automotive industries.

The Federal Trade Commission also estimated a Japanese cost advantage in auto production of $1,500–$2,000, this time for 1983, based primarily on its Bureau of Economics staff research (FTC, 1983, Appendix B). While the FTC criticises the AHH approach and findings, the FTC analysis is also subject to criticism. The FTC's estimates are obtained by projecting a 1980 estimate to December 1983, adjusting only the relative exchange rates between the two periods (1980: $1 US = 212 yen; 1983: $1 US = 240 yen). This simplistic approach cannot provide a reasonable estimate of 1983 unit cost differentials. The US automobile industry was operating at 58 per cent of capacity in 1980, and 80 per cent in 1983. Japanese auto producers utilised approximately 100 per cent of capacity during the entire 1980–1983 period. The relative increase in capacity utilisation for American firms by itself will narrow the Japanese cost advantage. The FTC also calculated the Japanese cost advantage for 1979 and 1980. According to the FTC's analysis, in 1979 only Toyota and Honda (in one of two cases) had a landed cost advantage over US producers, and these estimated advantages ($240 to $860) are *low* compared with the other estimates for the same year. The 1980 estimates show a substantial widening of the Japanese advantage. The FTC attributes this widening to three effects (p. 800):

(1) depreciation of the yen,
(2) recession in the US auto industry (underutilisation of capacity)
(3) relative increase in US labour costs due to increased relative wage rates and declining relative labour productivity.

The depreciation of the yen and an increased relative wage rate differential are *not* efficiency-related effects. Nor are they Japanese advantages which can be transferred through a JV.

The supposed increase in the Japanese labour-productivity advantage from 1979 to 1980 is also due to capacity utilisation effects. Capital structures, general equipment, special tools and overhead labour are fixed factors of production. Labour productivity in the US automobile industry fell by 6.4 per cent in 1979–80 while it increased in Japan by 3.5 per cent (FTC, pp. 801–2). Assuming there were no long-run labour productivity changes in US auto production in 1979–80, then the 6.4 per cent fall can be attributed to underutilisation of capacity. If the 3.5 per cent increase is a long-run Japanese efficiency improvement, if wages did not increase in Japan in 1979–80, if the average car cost $5,000 to produce in Japan and labour costs are 30 per cent of total costs, then labour productivity improvements increased the long-run cost advantage by only $53 between 1979 and 1980.

A SYNTHESIS

The above studies all place the Japanese cost advantage at $1,500–$2,000, but the apparently common bottom line is accidental. Our analysis of the studies indicated that they sometimes relied on erroneous data and always included short-run dislocations as part of productivity differentials.

Table 13.2 is our revision of the best of these studies. We have forced the total Ford and GM costs for an average American-sized automobile to equal $6,118 in 1979, which is the revised FTC 1979 unit cost estimates for average US production. The discrepancy was prorated across the categories as explained in detail in Fuss and Waverman (1987).

The Japanese cost advantage ranges from a high of $507 (Ford-Toyo Kogyo) to a low of $146 (GM-Nissan). Our revised Japanese cost advantage based on ACK data is nowhere near the $1,200–$1,500 cost advantage claimed by ACK (p. 63). However, it is reasonably close to the revised FTC estimates, when Toyota is not included in the FTC estimates.

We can extend the analysis to 1983 and decompose the Japanese cost advantage, given the above information (see Fuss and Waverman (1987), Chapter 2, for detailed calculations). We choose the highest cost differential (Ford-Toyo Kogyo). The decomposition, shown in Table 13.3, indicates wide variations in the advantages

Table 13.2: Comparative costs[a] and labour productivity in selected United States and Japanese automobile companies, 1981[b] (small car only)

Productivity/cost category	Ford	GM	Toyo Kogyo	Nissan
Labour productivity				
Employee hours per small car	67	67	51	50
Costs per small car				
Labour	$1,478	$1,480	$ 602	$ 576
Purchased materials	2,920	2,758	2,748	2,748
Other manufacturing and non-manufacturing costs	610	677	1,050	1,150
Equity costs	394	293	95	188
Tariff and transportation	0	0	400	400
TOTAL	$5,402	$5,208	$4,895	$5,062

Notes: a. Exchange rate used is 220 yen = $1 US.
b. As 1981 is a poor year for the US auto industry, the cyclically sensitive data (fixed and quasi-fixed factors), are taken from 1979 and extrapolated to 1981.
Source: Based on ACK (1983, p. 61, table 5.2).

Table 13.3: Decomposition of Japanese cost advantage 1983 (Ford-Toyo Kogyo)

	$
Cost advantage to Toyo Kogyo	915
Labour productivity effect (computed at average wage)	278
Wage effect (computed at average hours)	554
Purchased materials	−29
Other manufacturing costs and non-manufacturing costs	−517
Equity costs	292
Tariff and transportation costs	−429
Undervalued yen[a]	766

Note: a. The actual yen to dollar exchange rate of 240 is compared to its Fundamental Equilibrium Exchange Rate (FEER) of 205 (see Williamson, 1983).
Source: Based on Fuss and Waverman (1987).

of certain factors in the two countries. The American firms have an advantage in 'other manufacturing and non-manufacturing costs', purchased materials and 'tariffs and transportation'. The Japanese advantages are in labour productivity, wages, equity costs and the substantial effect of the undervalued yen.

The quantitatively most important sources of the 1983 Japanese cost advantage are first the undervalued yen and second the wage rate effect. This conclusion calls into question the ultimate benefits of transferring Japanese technology to North American automobile producers. *Japanese* (Tokyo Kogyo) *labour productivity* advantages, after accounting for other inputs and an undervalued exchange rate, amount to only *$278*, a far cry from the $1,500 estimates which the other studies call labour productivity advantages.

JOINT VENTURES

The existence of JVs and their potential for technology transfer are important issues. Undoubtedly, the growing number of JVs increase concentration in various national auto markets, signalling potential dangers to competition and potential rising prices. On the other hand, learning through JVs might make certain firms more viable competitors.

The maximum hypothetical cost savings of a Japanese-designed and run US assembly and stamping plant

What are the specific aspects of Japanese automotive production technology that could be transferred to US plants by these JVs? We have earlier outlined in general terms the potential sources of Japanese cost advantages in automobile production. This alleged Japanese cost advantage in automobile production is due to lower factor prices (lower wages, possible subsidised capital); differences in technology; a different organisation of technology; and, finally, a different work ethic in Japan (lifetime employment, superior labour management relations, intrinsically more devoted and harder-working labour). The 'technology' that could be transferred to the West would involve superior labour management, superior management of technology and any transferable technological differences. Lower wages, subsidised capital, an intrinsically harder-working labour force and cultural factors are elements of Japanese production which cannot be transferred to production in other countries.

The so-called 'efficiencies' must, therefore, be carefully examined to see if, first, they are transferable and, second, if they represent cost savings at North American factor prices.

Earlier portions of this paper have presented our views as to the possible unit cost advantage of Japanese auto producers (producing in Japan) over their American counterparts. Table 13.3 presents our best judgement as to the unit cost differential and its sources (based on aggregate accounting data and plant surveys as given in AHH and ACK) between Ford and Toyo Kogyo as of December 1983.

For our analysis of the cost savings inherent in technology transfer through JVs, we would have preferred to analyse directly a GM-Toyota cost comparison since these two firms have formed the first operating JV in the USA. However, a detailed breakdown for GM and Toyota such as is given for Ford-Toyo Kogyo (Mazda) is unavailable in the literature.[5]

The total estimated $915 cost advantage of Toyo Kogyo over Ford in 1983, shown in Table 13.3, is for *all* production activities — parts production, stamping and assembly. However, the Mazda US plant (like the GM-Toyota JV plant in Fremont, California) is not a fully-integrated auto production plant: assembly and stamping and some parts production will take place in the USA. Any technology transfer which is possible must be concentrated in assembly and stamping, not in engine or drivetrain production.

We estimate the hypothetical cost advantage *from US operations* of a Japanese firm operating an assembly and stamping plant in the USA in the following way, as presented in Table 13.4.

Recent analysis suggests that stamping and assembly involve one-third of auto production plant activities.[6] The first item in Table 13.4 assumes that 33 per cent, or about $90 of the $278 labour productivity advantage for Toyo Kogyo over Ford production shown in Table 13.3, occurs in the USA when assembly and stamping are switched there from Japan. Two potential wage effects are given in line 2. First, it is assumed that 33 per cent of the wage savings shown in Table 13.3 accrue to US production in the 'new' assembly and stamping plant (column A). This is contrary to our earlier arguments that lower factor prices in Japan cannot be transferred and represents wage concessions from US labour. Second, it is assumed that wage rates for a new Japanese plant in the USA are equivalent to wages in existing US plants so that no savings accrue to the US-based Japanese-run plant (a 'loss' to US assembly over assembly in Japan (column B)). The change in venue of assembly and stamping to the USA is assumed to have no impact on relative purchased materials costs (line 3). Therefore there is no change in the $29 estimated advantage to US production over Japanese production given in Table 13.3. In Table 13.3 the 'other manufacturing and non-manufacturing costs' (advertising, sales, administration, etc.) show a large disadvantage to Toyo Kogyo — $517. We assume that none of these disadvantages will be reduced simply

Table 13.4: Hypothetical sources of Japanese unit cost advantage: US assembly and stamping

	A. Average wages		B. US wages	
	US	Japan	US	Japan
1. Labour productivity	90	189	90	189
2. Wage effect	180	374	0	374
3. Purchased materials		−29		−29
4. Other manufacturing and non-manufacturing costs	0	−517	0	−517
5. Annual unit equity costs	73	146	37	146
6. Tariff and transportation	162	−429	162	−429
7. Undervalued yen	0	313	0	313
	505	47	289	47
	$552		$336	

1. 33 per cent of value in Table 13.3 attributed to US production.
2. Two cases: A. 33 per cent of wage differential in Table 13.3; B. no savings.
3,4. No differences from values in Table 13.3.
5. Assume that half the capital is in place in US (assembly and stamping are more capital intensive than engine and transmission production), that the Japanese can economise on stamping and assembly plant costs and therefore save 25 per cent (B) to 50 per cent (A) of the annual unit equity cost savings from Japan-sourced plants.
6. The tariff is calculated as 2.9 per cent of $2,000 worth of parts imports, a savings of $87 over the tariff cost of importing a $5,000 wholesale-valued car. Transportation costs savings are estimated at $75 over shipping the whole car from Japan.
7. Estimated as $200 × (220/205 − 220/240). Alternatively, one can think of US stamping and assembly as a $453 penalty ($766 − 313).

Source: Based on AHH (1981) and ACK (1983).

by shifting assembly and stamping to the USA and so there is no difference in this item (4) between Tables 13.3 and 13.4.

The equity cost calculations are as follows. We assume that stamping and assembly are more capital intensive than parts production so that one-half of the capital of an integrated automobile plant would be in the US assembly and stamping plant and half the capital in the Japanese parts production plants. The Japanese are assumed to economise on capital costs in the USA as compared to existing US producers and so save 25 per cent (B) to 50 per cent (A) of the Japanese advantages in unit capital costs shown in Table 13.3.[7] As a result, Japanese production of parts 'costs' $146 less in equity costs than if the parts were produced in the USA (½ of the $292 shown in line 5 of Table 13.3) and the JV's stamping and assembly plants save $37 to $73 in equity costs as compared to a US-owned and run plant.

The tariff and transportation cost calculations are detailed in line 6 of Table 13.4. The yen-dollar effect (item 7) is estimated by assuming that $2,000 of components are imported at an exchange rate of 240 yen/dollar instead of at the FEER of 205 yen/dollar.

In total, this analysis indicates a hypothetical maximum advantage of $336 to $552 for Japanese (Mazda) over US (Ford) production from a US-based JV assembly and stamping plant and Japanese sourcing of major components. Importing the finished car directly from Japan led to a $915 Japanese advantage (Table 13.3), thus shifting stamping and assembly to the USA incurs a unit cost penalty of $363 to $579 per car.[8]

The US-based, Japanese designed and run stamping and assembly plants are estimated as having hypothetical maximum lower unit costs as compared to US producers of between $336 and $552 per car. However, to analyse the savings transferable in a JV we ignore the tariff, transportation and yen/dollar components of our decomposition. These latter components, while affecting the decision on where to locate from the viewpoint of the Japanese firm are not, however, savings transferable through knowledge transfer.

As a result, the *maximum* real savings transferable through a JV to US producers is $127 ($90 + $37) to $343 ($90 + $180 + $73).

What portions of the Japanese technology are transferable to North America?

We have now estimated that a Japanese stamping and assembly plant in the US or a JV of the GM-Toyota or Chrysler-Mitsubishi type based on these particular data and assumptions could save, at most, $127 to $343 per car as compared to an existing US producer's plant costs.[9] Acknowledging this maximum hypothetical cost differential is not equivalent to arguing either that a JV is necessary for new US plants to reduce costs by these amounts or that this hypothetical cost saving would indeed be transferred by a JV to the US.

A more detailed look at the differences in automobile production technology is warranted at this point. The superior technology of Japanese auto firms is said to be based on (a) design and manufacturing, (b) layers of management, (c) job classifications, (d) worker training, (e) productivity, (f) material scheduling and (g) quality.[10] As much has been written on these differences, we will only highlight the issues concentrating on identifying those elements that could be transferred out of Japan.

The Japanese auto firms do not appear to use a very different technology from US firms. While the Japanese had more mechanised assembly lines, this gap appears to have decreased. What is clearly different is the method of production. The Japanese use a 'just-in-time' production and inventory control, a Kanban system for inventories, in-line scheduling, in-line quality control, mass relief, minimal job classifications, and quality circles where all workers are involved in quality control and productivity improvements.

A typical Japanese assembly plant runs as follows. Cars are produced to demand; the assembly line can accommodate different models (e.g. two door, four door) of the same wheelbase. Cars follow in-line sequencing for different engines, colours, etc., never leaving their place in line. Parts including preassembled seats are delivered by their manufacturers 'just-in-time' (say four to twelve hours) before being required on the line. Die changes are extraordinarily quick (in some cases 15 minutes). The entire line of workers is relieved at once. Five or six job classifications exist so that workers do several jobs. The line attention to quality is extreme; a line shuts down if a worker notices a defect he cannot correct. All workers are asked to contribute to design, quality and productivity improvements.

Contrast an older Detroit auto assembly plant. The assembly line is physically far removed from stamping and parts production. As a result, large inventories of parts are held at the assembly plants. Production of cars is for inventory as well, with batches of identical cars produced and held for demand. Dies are changed slowly (say several hours) for production of a new model. Repairs are made to cars at the end of the process. Numerous job classifications (over 100) abound; assembly line workers are never involved in vehicle design. It is anathema to stop the line, so that defects continue to the end of the line.

Most of the above technological nuances are well known to US firms. The Kanban system was developed in the US. Quick die changes are also well known in other American industries. US firms have begun to utilise in-line sequencing in new automobile plants, and to reduce the number of job classifications. In short, some aspects of the Japanese technology are already being used in the USA.

A number of hypotheses can be advanced to explain the differences in technological development. Some observers (Saxonhouse (1985), Gomez-Ibanez and Harrison (1982)) suggest that observed differences in auto production techniques between the USA and

Japan are primarily due to differences in relative factor prices. Japan is a small country with relatively high land costs. As a result, it is argued, Japanese production economises on land; witness the smaller Japanese auto plants, in-line sequencing, and the Kanban system, all designed to reduce land usage. For a large country like the USA with relatively low land costs, these aspects of the Japanese technology may be 'inefficient'. The differences in US and Japanese technology could represent different points on an isoquant (the differences being due to different factor price ratios), rather than representing a more efficient use of factors in Japan. If this were true then the Japanese technology represents a response to factor prices which are different from those in the USA. American firms and Japanese firms operating in the USA then would not use the Japanese technology at US factor prices. This observation, while important and largely ignored in the numerous volumes extolling Japanese production systems, cannot explain lower *total* costs for Japanese production compared to US production.

If there is some truth to the relative factor price story, what explains the move to Japanese-style production in the US auto plants? Some observers (e.g. Crandall, 1984) suggest that until the 1970s the North American auto industry was essentially protected by differences in consumer tastes — US consumers desired much larger cars than consumers in the rest of the world. This protection led to an oligopoly which over time shared rents with labour. Auto workers took high pay and protection in terms of job classifications. Quality of output fell. The enormous changes in oil prices fuelled changes in North American consumers' taste for cars, opening the North American market to foreign producers. It has taken these North American firms many years to adapt to this new competition. Part of this story suggests that JVs and Japanese production in the US are important vehicles which can be used to reduce the rents to US labour by pressuring the auto union to lower real wages and reduce the number of job classifications. This story, while perhaps an important ingredient in explaining JVs in the US with Japanese firms, cannot explain why these JVs use Japanese production techniques.

Therefore, while we feel that there is a certain amount of truth to these suppositions (relative factor prices, union busting), they do not fully explain the facts. The different organisation of auto production in Japan than in the USA — the Kanban, smaller plants, the greater integration of functions within one facility — might have originated from the pressure of relative factor prices in Japan.

However, at this point, these practices also appear to represent a more efficient technology at US factor prices. What is unclear to us is why a JV is necessary to transfer much of this technology.

In Table 13.4 we estimated a hypothetical maximum value of \$127 to \$343 for the gains from Japanese assembly and stamping in the USA over US production, were the maximum differential between Mazda and Ford costs to hold. How much of this cost reduction could US firms produce without a Japanese partner? The data in Table 13.4 refers to *past practice* data — comparing existing Mazda Japanese plants with existing Ford US plants. Newer North American plants can, however, install some portions of the Japanese technology without a JV with a Japanese firm.[11] Newer North American plants have already greatly reduced the number of job classifications. In-line sequencing, Kanban, quality circles (workers taking part in productivity improvements), mass relief are used in newer US plants.[12] However, while these plants have lower costs than older plants, none of the US firms argue that they can now meet Japanese costs of production.

But how could these US plants meet Japanese *costs of production* where there is presently a large labour cost advantage and no productivity disadvantage (perhaps an advantage) to Japan? North American automobile producers, in time, can compete with Japanese producers. First, since relative factor prices differ between North America and Japan, greater substitution for labour in North America than in Japan may be warranted.[13] Second, the tariff and transportation costs act as a cost advantage to US firms. Third, there may be no intrinsic reason why labour productivity (as we have defined it) in the USA should not be as high as in Japan. If however, there is some fundamental cultural difference between Japanese and US workers — the Japanese just work harder at all real wages and technology does not allow sufficient substitution away from labour — then there is a real productivity dilemma. Only a decline in the relative real wages of US automobile workers or increased protection could maintain the US auto industry as competitors to the Japanese. But, no JV can transfer this work ethic to the USA.

In short, much of the Japanese productivity (not cost) advantage appears to be transferable without a JV. The most important function of the JV may be to rewrite contracts with labour thus reducing the Japanese wage advantage, the major source of the Japanese cost advantage.

SUMMARY

In this paper we have shown that previous estimates of the productivity advantages of Japanese auto producers of some $1,500 to $2,000 over US auto producers were severely exaggerated.

(1) The numbers are more in the $600 to $900 range than in the $1,500 to $2,000 range.
(2) These estimates represent an aggregation of a large number of disparate effects:
   (a) differing factor prices,
   (b) differing rates of capacity utilisation,
   (c) exchange rates differing from their fundamental equilibrium ratios,
   (d) differing technology,
   (e) differing labour productivity.

We have argued that labour productivity in the two countries can only be examined by holding all other factors constant and that it is these other factors (principally wage rates and the yen/dollar ratio) which account for most of the Japanese cost advantage in auto production.

We then examined a hypothetical Japanese-designed and run US based assembly and stamping plant. We estimated that the hypothetical maximum cost advantage for such a plant over an existing US facility was $127 to $343. Not all of this 'advantage' could be transferred with a JV nor is a JV the only means of transferring part of the cost advantage. The major advantage that Japanese firms producing in the USA may have over US firms are their different (and lower cost) labour contracts.

NOTES

1. The authors were consultants to Chrysler.

2. Ford owns 24 per cent of Mazda; Chrysler 20 per cent of Mitsubishi.

3. GM has equity shares in two Japanese producers — 34 per cent of Isuzu and 5 per cent of Suzuki (as well as 50 per cent of the Korean firm Daewoo). GM argued that Isuzu was too financially weak and Suzuki too unknowledgeable to undertake the kind of JV necessary in the US market.

4. Fuss and Waverman (1987), calculate that in 1974 the Japanese automobile industry actually had a total cost *disadvantage* of 9 per cent. The

difference in estimates arises from two sources. First, Toder *et al.* assume auto production is equally efficient in the two countries, whereas Fuss and Waverman estimate that, in 1974, auto production in the US was more efficient. Second, Toder *et al.* assume equal capital costs, whereas Fuss and Waverman estimate that the US had a cost of capital advantage.

5. Toyota is the most efficient Japanese producer and the Toyo Kogyo costs probably overestimate Toyota's costs. However, GM is the most efficient US car producer and it is therefore unclear as to whether the cost differential between GM and Toyota would be more or less than between Toyo Kogyo and Ford.

6. Kwoka (1983).

7. There is no 'transfer' of subsidised capital costs (if they exist) from Japan.

8. This may explain Japanese reluctance to invest in the USA until the Voluntary Restraint Agreement imposed by the US government in April 1981.

9. Ignoring any wage effect or assuming that wage concessions will occur for all firms, US or foreign, suggests unit cost savings of $127 to $163.

10. See FTC (1983), pp. 676–93.

11. See reports of new Ford and Chrysler plants in *Automotive News*, June to December 1985.

12. Chrysler's Sterling Heights, Michigan, plant has fewer than ten job classifications.

13. We are beginning to see *more* automated plants in the USA than in Japan.

Bibliography

Abernathy, W.J., Harbour, J.E. and Henn, J.M. (1981) *Productivity and Comparative Cost Advantages: Some Estimates For Major Automobile Producers,* draft report to the US Department of Transportation, Transportation Systems Center (February)

———, Clark, K.B. and Kantrow, A.M. (1983) *Industrial Renaissance,* New York, N.Y.: Basic Books

Amsalem, M.A. (1982) 'Management: The Forgotten Factor in Technology Choice', in M. Srinivasan (ed.), *Technology Assessment and Development,* New York: Praeger

Aquino, A. (1978) 'Intra-Industry Trade and Inter-Industry Specialization as Concurrent Sources of International Trade in Manufactures', *Weltwirtschaftliches Archiv Bd., CXIX,* 90–9

AREPIT (Association de Recherche Economique en Propriété Intellectuelle et Transferts Techniques) (1985) *The Role of Patents in Multinational Corporation Strategies,* Paris

Armington, P. (1969) 'A Theory of Demand for Products Distinguished by Place of Production', IMF, *Staff Papers* (March), 159–78

Arpan, Jeff S. and Ricks, David (1974) 'Foreign Direct Investment in the United States and Some Attendant Research Problems', *Journal of International Business Studies, 5,* 1–7

Astwood, D.M. (1981) 'Canada's Merchandise Trade Record and International Competitiveness in Manufacturing, 1960 to 1979', in K.C. Ohawan, H. Etemad and R.W. Wright (eds.), *International Business: A Canadian Perspective,* Don Mills, Ontario: Addison-Wesley, 48–73

Bain, J. (1966) *International Differences in Industrial Structure,* Cambridge: Harvard University Press

Baldwin, J.R. and Gorecki, P.K. (1983a) *The Relationship Between Plant Scale and Product Diversity in Canadian Manufacturing Industries,* Ottawa: Economic Council of Canada

——— (1983b) *Trade, Tariffs, Product Diversity and Length of Run in Canadian Manufacturing Industries: 1970–1979,* Ottawa: Economic Council of Canada

——— (1983c) *Trade, Tariffs and Relative Plant Scale in Canadian Manufacturing Industries: 1970–1979,* Ottawa: Economic Council of Canada

Baranson, J. (1978) *Technology and the Multinationals: Corporate Strategies in a Changing World Economy,* Lexington, Mass.: Lexington Books

Becker, G. (1964), *Human Capital,* NBER, New York

Behrman, J.N. (1958) 'Licensing Abroad Under Patents, Trademarks and Know-how by United States Companies', *The Patent, Trademark and Copyright Journal of Research and Education* (June)

——— (1960) 'Promoting Free World Economic Development Through Direct Investment', *American Economic Review* (May)

——— and Fischer, W.A. (1979) 'The Coordination of Foreign R&D

Activities by Transnational Corporations, *Journal of International Business Studies, 10,* 3 (Winter), 28–35

——— (1980a) *Overseas R&D Activities of Transnational Companies,* Oelgeschlager, Cambridge, Mass.: Gunn and Hain Publishers

——— (1980b) 'Transnational Corporations: Market Orientations and R&D Abroad', *Columbia Journal of World Business, 15,* 3 (Fall), 55–60

——— and Wallender, H.W. (1976) *Transfers of Manufacturing Technology Within Multinational Enterprises,* Cambridge, Mass.: Ballinger Publishing

Bhagwati, J. (1964) 'The Pure Theory of International Trade: A Survey', *Economic Journal* (March), 1–84

——— (1972) Book Review: 'Raymond Verson, Sovereignty at Bay: the Multinational Spread of US Enterprises, 1971', *Journal of International Economics* (September)

Bishop, P. and Crookell, H. 'Specialization and Foreign Investment in Canada', in D.G. McFetridge (ed.), *Canadian Industry in Transition,* Ottawa: Royal Commission on the Economic Union

Blomstrom, M. (1984) *The Forms of Foreign Involvement by Swedish Multinationals,* Paris: OECD

Bonin, B. (1984) *L'entreprise multinationale et l'Etat,* Montreal: Editions Etudes Vivantes

Boretsky, M. (1982) 'The Threat of US High Technology Industries: Economic and National Security Implications', International Trade Administration, US Department of Commerce (March)

Brean, Donald J.S. (1984) *International Issues in Taxation: The Canadian Perspective*, Toronto: Canadian Tax Foundation

Britton, J.M.H. and Gilmour, J.M. (1978) *The Weakest Link, A Technological Perspective on Canadian Industrial Underdevelopment,* Ottawa: Minister of Supply and Services

Buckley, P. (1979) 'The Modern Theory of the Multinational Enterprise', *Management Bibliographies and Reviews,* 171–85

——— and Davies, H. (1981) 'Foreign Licensing In Overseas Operations: Theory and Evidence from the UK', in R.G. Hawkins and A.J. Prasad (eds.), *Technology Transfer and Economic Development,* Greenwich, Conn.: JAI Press

Business International S.A. (1971) *European Business Strategies in the United States, Meeting the Challenge of the World's Largest Market,* Business International, Geneva (September)

Cacnis, D.G. (1985) 'The Sources of Total Factor Productivity Growth in Canada, 1950–1976', New York: Adelphi University, Mimeograph

Casson, M. (1979) *Alternatives to the Multinational Enterprise,* New York: Holmes and Meier Publishers Inc.

Caves, R.E. (1975) *Diversification, Foreign Investment and Scale in North American Manufacturing Industries,* Ottawa: Economic Council of Canada

——— (1982) *Multinational Enterprise and Economic Analysis,* Cambridge: Cambridge University Press

——— (1983) 'Multinational Enterprises and Technology Transfer', in A.M. Rugman (ed.), *New Theories of the Multinational Enterprise,* New York: St. Martin's Press

———, Crookell, H. and Killing, P. (1982) 'The Imperfect Market for Technology Licenses', Discussion Paper No. 903, Harvard Institute of Economic Research, Harvard University

Commission of Inquiry on the Pharmaceutical Industry (1985) *Report*, Ottawa: Supply and Services Canada

Contractor, F.J. (1981) *International Technology Licensing*, Lexington, Mass.: Lexington Books

——— (1981a) 'The Role of Licensing in International Strategy', *Columbia Journal of World Business* (Winter)

——— (1983b) 'Technology Importation Policies in Developing Countries: Some Implications of Recent Theoretical and Empirical Evidence', *Journal of Developing Areas, 17* (July), 499–520

——— and Sagafi-Nejad, T. (1981b) 'International Technology Transfer: Major Studies and Policy Responses', *Journal of International Business Studies* (Fall)

——— (1983a) 'Technology Licensing Practice in US Companies: Corporate and Public Policy Implications', *Columbia Journal of World Business* (Fall)

Cory, P.F. (1983) 'Industrial Co-operation, Joint Ventures and the ME in Yugoslavia', in A.M. Rugman (ed.), *New Theories of the Multinational Enterprise*, New York: St. Martin's Press

Couffin, H. (1977), 'Les entreprises francaises sur le marché américain', *Economica*, Paris, 44–8

Coughlin, C.C. (1983) 'The Relationship Between Foreign Ownership and Technology Transfer', *Journal of Comparative Economics, 7*, 400–14

Crandall, R.W. (1984) 'Import Quotas and the Automobile Industry: The Costs of Protectionism', *The Brookings Review, 2* (November), 8–16

Creamer, D. (1976) *Overseas Research and Development by United States Multinationals, 1966–1975: Estimates of Expenditures and a Statistical Profile*, The Conference Board, New York

Crookell, H. (1973) *The Transmission of Technology Across National Boundaries*, Ottawa: Department of Industry, Trade and Commerce

———, Caves, R.E. and Killing, J.P. (1984) *Getting Along Without Multinational Firms*, unpublished paper, The University of Western Ontario, London, Ontario

Daly, D.J. (1976) 'Canadian Management: Past Recruitment Practices and Future Training Needs', in Max von Zur-Muehlen (ed.), *Highlights and Background Studies*, Ottawa: Canadian Federation of Deans of Management and Administrative Studies, 178–200

——— (1984) 'Technology Transfer and Canada's Competitive Performance', Downsview: York University Mimeograph, for Third Annual Workshop on US-Canadian Relations, Ann Arbor, Michigan

——— and Globerman, S. (1976) *Tariff and Science Policies: Applications of a Model of Nationalism*, Toronto: University of Toronto Press

———, Keyes, B.A. and Spence, E.J. (1968) *Scale and Specialization in Canadian Manufacturing*, Ottawa: Economic Council of Canada

——— and MacCharles, D.C. (1986a) *Canadian Manufacturing: International Performance and Policy Options*, Vancouver: Fraser Institute, forthcoming

——— (1986b) *Canadian Manufactured Exports: Constraints and Opportunities,* Montreal: Institute for Research on Public Policy, forthcoming

——— and Altwasser, W. (1982) 'Corporate Profit Drop Worst Since 1930s', *Canadian Business Review,* Ottawa: Conference Board of Canada, *9* (3), 6–12

Davidson, W.H. (1980) *Experience Effects in International Investment and Technology Transfer,* Ann Arbor: UMI Research Press

——— (1982) 'Trends in Transfer of US Technology to Canada', in *The Adoption of Foreign Technology by Canadian Industry,* Science Council of Canada, Ottawa

——— and Harrigan, R. (1977) 'Key Decisions in International Marketing: Introducing New Products Abroad', *The Columbia Journal of World Business* (Winter)

——— and McFetridge, D.G. (1984) 'International Technology Transactions and the Theory of the Firm', *Journal of Industrial Economics* (March), 253–64

——— (1985) 'Key Characteristics in the Choice of International Technology Transfer Mode', *Journal of International Business Studies* (Summer)

Davies, H. (1977) 'Technology Transfer Through Commercial Transactions', *Journal of Industrial Economics* (December)

De Bodinat, H. (1984) 'Influence in the Multinational Enterprise: The Case of Manufacturing', in R. Stobaugh and L.T. Wells Jr. (eds.), *Technology Crossing Borders: The Choice, Transfer, and Management of International Technology Flows,* Harvard Business School Press, Boston, 265–92

Delapierre, M. and Michalet, C.A. (1984) *Les 'nouvelles formes' d'investissement dans les pays en voie de developpement: le cas francais,* Paris: OECD

DeMelto, D.P., McMullen, K.E. and Wills, R.M. (1980) *Preliminary Report: Innovation and Technological Change in Five Canadian Industries,* Ottawa: Economic Council of Canada

Dreze, J. (1960) 'Quelques reflexions sereines sur l'adaptation de l'industries belge au Marché Commun', *Comptes-rendus des travaux de la Societe Royale d'Economie Politique de Belgique* (December)

Duerr, M.G. (1970) 'R&D in the Multinational Company, A Survey', *Managing International Business, no. 8,* New York: The Conference Board

Dunn, Robert M. Jr. (1978) *The Canadian-US Capital Market,* Toronto: C.D. Howe Institute

Dunning, J.H. (1979) 'Explaining Changing Patterns of International Production: In Defence of the Eclectic Theory', *Oxford Bulletin of Economics and Statistics* (November), 269–95

——— (1981) 'Explaining the International Direct Investment Position of Countries: Towards a Dynamic or Developmental Approach', *International Production and the Multinational Enterprise*, London: George Allen & Unwin, 109–41

Dunning, John H. and Rugman, Alan M. (1985) 'The Influence of Hymer's Dissertation on the Theory of Foreign Direct Investment', *American Economic Review* Papers, *75,* 2, May, 228–32

Eastman, H. (1982), 'The Objectives and Structures of Canadian Multinational Enterprises', paper prepared for the Canadian Economics Association, Annual Meeting (December), mimeo

Economic Council of Canada (1983) *The Bottom Line,* Ottawa

Economist (1984) 'South Korean Electronics, Ramming the Japanese', (18 February)

Erdilek, Asim (1985) *Multinationals as Mutual Invaders: Intra-Industry Direct Foreign Investment,* London: Croom Helm

Etemad, H. and Séguin Dulude, L. (1984) *R&D and Patenting Characteristics of Canadian World Product Mandated Subsidiaries: Some Theoretical Discussions and Empirical Evidence,* Les Cahiers du CETAI no. 84–14, Centre d'études en administration internationale, Ecole des Hautes Etudes Commerciales (December)

——— (1985) 'R&D and Patenting Patterns in 25 Large MNEs', *Proceedings of the Annual Conference of the Administrative Sciences Association of Canada, International Business Division (ASAC), 6,* part 8, 21–32

European Management Forum (1984 and 1985) *Annual Report*

Evenson, R.E. (1984) 'International Invention: Implications for Technology Market Analysis', in Z. Griliches (ed.), *R&D, Patents and Productivity,* Chicago: University of Chicago Press

Federal Trade Commission (1983) *Report of the Bureaus of Competition and Economics Concerning the General Motors/Toyota Joint Venture,* 3 volumes, mimeo, Washington, DC

Flamm, K. and Pelzman, J. (1984) *New Forms of Investment by US Firms in Emerging and Declining Sectors: Textiles and Microelectronics,* Paris: OECD

Franko, L. (1971) *European Business Strategies in the United States,* Business International SA, Geneva

——— (1976) *The European Multinationals, a Renewed Challenge to American and British Big Business,* Stamford: Greylock Publishers

——— (1984) *Practices of Selected US Companies in the Automobile, Pharmaceuticals and Fine Chemicals, Food Processing and Mining Industries,* Paris: OECD

Fuss, M. and Waverman, L. (1985) 'Productivity Growth in the Automobile Industry, 1970–1980: A Comparison of Canada, Japan and the United States', NBER Working Paper No. 1735, Cambridge, Mass.

——— (1986a) 'The Extent and Sources of Cost Efficiency Difference Between US and Japanese Automobile Producers', NBER Working Paper No. 1849, Cambridge, Mass.

——— (1986b) 'The Canada-US Auto Pact of 1965: An Experiment in Selective Trade Liberalization', International Economics Program Working Paper No. DP 86–6, University of Toronto, Toronto, Ontario

——— (1987) *Costs and Productivity in Automobile Production,* New York, NY: Cambridge University Press

Globerman, Steven S. (1979) *US Ownership of Firms in Canada: Issues and Policy Approaches,* Toronto: C.D. Howe Institute

——— (1979) 'Foreign Direct Investment and "Spillover" Efficiency Benefits in Canadian Manufacturing Industries', *Canadian Journal of Economics, 12,* 42–56

——— (1985) 'Direct Investment, Economic Structure, and Industrial Competitiveness: The Canadian Case', in John H. Dunning (ed.), *Multinational Enterprises, Economic Structure and Industrial Competitiveness,* London: John Wiley and Sons

Gomez-Ibanez, J.A. and Harrison, D. (1982) 'Imports and the Future of the US Automobile Industry', *American Economic Review* (May), 317–23

Government of Canada (1983) *Manufacturing Trade and Measures 1966–1982: Tabulations of Trade, Output, Canadian Market, Total Demand and Related Measures for Manufacturing Industrial Sectors,* Ottawa: Department of Industry, Trade and Commerce/Regional Economic Expansion

Graham, F.D. (1925) 'Some Aspects of Protection Further Considered', *Quarterly Journal of Economics* (February), 199–227

Gray, Herb (1972) *Foreign Direct Investment in Canada,* Ottawa: Information Canada

Gray, Peter (ed.) (1985) *Uncle Sam as Host: Foreign Direct Investment in the United States,* Greenwich, Conn.: JAI Press

Grubel, H.B. and Scott, A.D. (1966) 'The International Flow of Human Capital', *American Economic Review* (May), 268–74

Guisinger, S. and Associates (1985) *Investment Incentives and Performance Requirements,* New York: Praeger

Hakansson, H. and Laage-Hellman, H. (1984) 'Developing a Network R&D Strategy', *Journal of Product Innovation Management, 1,* no. 4 (December), 224–37

Hanel, P. and Palda K. (1981) *Innovation and Export Performance in Canadian Manufacturing,* Ottawa: Economic Council of Canada (December)

Harris, R.G. with Cox, D. (1983) *Trade, Industrial Policy, and Canadian Manufacturing,* Toronto: Ontario Economic Council

Helleiner, G.K. (1979) 'Transnational Corporations and Trade Structure: The Role of Intra-Firm Trade', in H. Giersch (ed.), *On the Economics of Intra-Industry Trade,* Tubingen, Germany: J.B. Mohr

——— and Lavergne, R. (1979) 'Intra-Firm Trade and Industrial Exports to the United States', *Oxford Bulletin of Economics and Statistics,* XLI (4), 297–311

Hewitt, G. (1980) 'Research and Development Performed Abroad by US Manufacturing Multinationals', *Kyklos, 33,* fasc. 2, 308–27

Hirsch, S. (1976) 'An International Trade and Investment Theory of the Firm', *Oxford Economic Papers* (July), 258–69

Hufbauer, G. (1966) *Synthetic Materials and the Theory of International Trade,* Cambridge: Harvard University Press

Industry Studies Group (1979) 'US Industrial R&D Spending Abroad', *Reviews of Data on Science Resources, 33,* National Science Foundation, Washington, DC, NSF 79–304 (April), 1–7

International Trade Administration (1983) 'An Assessment of US Competitiveness in High Technology Industry', US Department of Commerce

Jenkins, G.P. (1985) 'Options for Dealing with Declining Industries', Ottawa: Papers on the Issues Facing the Conference, National Economic Conference

Jones, R.W. (1984) 'Protection and the Harmful Effects of Endogenous Capital Flows', *Economic Letters*

Katrak, H. (1974) 'Human Skills, R and D and Scale Economies in the Exports of the United Kingdom and the United States', *Oxford Economic Papers* (November), 337–60

Kelly, R.K. (1977) 'The Impact of Technological Innovation on International Trade Patterns', *Staff Economic Report*, Office of Economic Research, US Department of Commerce (December)

Killing, J.P. (1975) *Manufacturing Under Licence in Canada*, University of Western Ontario, London, Ontario

——— (1980) 'Technology Acquisition: Licence Agreement or Joint Venture', *The Columbia Journal of World Business* (Fall)

Kojima, K. (1978) *Direct Foreign Investment: A Japanese Model of Multinational Business Operations*, London: Croom Helm

Kopits, G.F. (1976) 'Intra-Firm Royalties Crossing Frontiers and Transfer Pricing Behaviour', *The Economic Journal* (December)

Kremp, Elisabeth and Larroumets, Valerie (1985) 'Les échanges internationaux de produits à haute technologies', *Economie Prospective Internationale, 23*, 7–34

Krueger, A.O. (1977) *Growth, Distortions and Patterns of Trade Among Many Countries*, Princeton Studies in International Finance, no. 40, Princeton University

Kwoka, J. (1983) Appendix in Federal Trade Commission (1983). *Report of the Bureaus of Competition and Economics Concerning the General Motors/Toyota Joint Venture*, 3 volumes, mimeo, Washington, DC

Lafleur, B. (1984) 'Forecasters See Slower Pace of Economic Recovery', *The Canadian Business Review*, Ottawa: Conference Board of Canada (Summer), 2–5

Lake, A.W. (1979) 'Technology Creation and Technology Transfer by Multinational Firms', in R.G. Hawkins (ed.), *Research in International Business and Finance, 1*, 137–87, Greenwich, Conn.: JAI Press

Lancaster, K. (1980) 'Intra-Industry Trade Under Perfect Monopolistic Competition', *Journal of International Economics*, 151–75

Lassudrie-Duchene, B. (1971) 'La demande de différence et l'échange international', *Economies et Societes*, Cahiers de l'ISEA (June), 961–82

Linder, S.B. (1961) *An Essay on Trade and Transformation*, New York: John Wiley and Sons

Litvak, I.A. and Maule, C.J. (1981) *The Canadian Multinationals*, Toronto: Butterworth and Company

MacCharles, D.C. (1978) *The Cost of Administrative Organizations in Canadian Manufacturing Industries*, Toronto: University of Toronto Dissertation

——— (1981) *The Performance of Direct Investment in the Manufacturing Sector*, Saint John, N.B.: University of New Brunswick, Mimeograph

——— (1982) *Summary of Ownership and Performance Differences: Non-Production Costs and Manufacturing Productivity*, Saint John, N.B.: University of New Brunswick, Mimeograph

——— (1983) 'Knowledge, Productivity and Industrial Policy', *Cost and Management*, Hamilton, Ontario: Society of Management Accountants

(January–February)
——— (1984) *Canadian Domestic and International Intra-Industry Trade*, Saint John, N.B.: University of New Brunswick
——— (1986) 'Canadian International Intra-Industry Trade', in P.K.M. Tharakan and D. Greenaway (eds.), *Intra-Industry Trade*, Sussex: Wheatsheaf Books, forthcoming
Machlup, F. (1962) *The Production and Distribution of Knowledge in the United States*, Princeton, N.J.: Princeton University Press
Maddala, G.S. (1983) *Limited Dependent and Qualitative Variables in Econometrics*, Cambridge: Cambridge University Press
Madeuf, Bernadette (1984) 'International Technology Transfers and International Technology Payments', *Research Policy* (June) *13*, 125–40
Magee, S.P. (1976) *International Trade and Distortions in Factor Markets*, New York: Marcel Dekker Inc.
Mansfield, Edwin *et al.* (1982) *Technology Transfer, Productivity and Economic Policy*, New York: Norton
——— (1984) 'R&D and Innovation: Some Empirical Findings', in Z. Griliches (ed.), *R&D, Patents and Productivity*, Chicago: University of Chicago Press, 127–48
——— and Romeo, A. (1979) 'Overseas Research and Development by US-based Firms', *Economica*, *46* (May), 187–96
——— (1980) 'Technology Transfer to Overseas Subsidiaries by US-based Firms', *The Quarterly Journal of Economics* (December)
——— and Wagner, S. (1979) 'Foreign Trade and US Research and Development', *The Review of Economics and Statistics*
Mathewson, G.F. and Quirin, G.D. (1979) *Fiscal Transfer Pricing in Multinational Corporations*, Ontario Economic Council and The University of Toronto Press, Toronto
McFetridge, D.G. (1985) 'The Economics of Industrial Policy: An Overview', in D.G. McFetridge (ed.), *Canadian Industrial Policy in Action*, Ottawa: Royal Commission on the Economic Union
——— and R.J. Corvari (1985) 'Technology Diffusion: A Survey of Canadian Evidence and Public Policy Issues', in D.G. McFetridge (ed.), *Technical Change in Canadian Industry*, Ottawa: Royal Commission on the Economic Union
McMullen, K. (1982) *A Model of Lag Lengths for Innovation Adopted by Canadian Firms*, Discussion Paper #216, Economic Council of Canada, Ottawa
Messerlin, P.A. (1984) *Commerce exterieur, risque et politique commerciale*, Rapport pour le Commissariat Général du Plan, Paris: mimeo.
Mucchielli, J.L. (1985) *Les firmes multinationales: mutations et nouvelles perspectives*, Paris: Economica
——— and Sollogoub, M. (1980) *L'échange international, fondements théoriques et analyses empiriques*, Paris: Economica
Mundell, R.A. (1957) 'International Trade and Factor Mobility', *American Economic Review*, *XLVII*, 321–35
Mytelka, L.K. (1978) 'Licensing and Technology Dependence in the Andean Group', *World Development* (April)
Nalin, C. and Delfava, G. (1984) *Implantation de Lafarge-Coppee sur le marché Nord Américain*, Master's degree thesis, Faculte d'Economie

Appliquée, Aix-Marseille III (June)
Nasbeth, L. and Ray G. (1974) (eds.), *The Diffusion of New Industrial Processes*, London: Cambridge University Press
Neary, J.P. and Ruane, F.P. (1985) 'International Capital Mobility, Shadow Prices and the Cost of Protection', Centre for Economic Policy Research, Discussion Paper 58
OECD (1981) Manuel de Frascati
OECD (1983) 'Etudes experimentales concernant l'analyse de l'output', Partie 2, le commerce international des produits de haute technologie
OECD (1984) La Balance des Paiements Technologiques, France DSTI/SPR/84, 38/08
OECD (1985) Les indicateurs de la science et de la technologie, SPT, 8
Ohlin, B. (1931) 'Protection and Non-Competing Groups', *Weltwirtschaftliches Archiv, Heft 1*, 30–45
——— (1968) *Interregional and International Trade*, Harvard University Press, 2nd edition
Oman, C. (1984) *New Forms of International Investment in Developing Countries*, Paris: OECD
Onida, F. *et al.* (1984) *New Forms of International Technology Transfer by Italian Enterprises to Developing Countries: A First Assessment*, Paris: OECD
Ozawa, T. (1984) *New Forms of Investment by Japanese Firms*, Paris: OECD
Palda, K.S. (1984) *Industrial Innovation, Its Place in the Public Policy Agenda*, Vancouver: Fraser Institute
Paribas (1984) 'Echange international et effort technologique', *Bulletin economique mensuel* (December)
Parry, T.G. (1984) *New Forms of International Involvement: Australian Companies in the Asia-Pacific Region*, Paris: OECD
Pattison, J.C. (1978) *Financial Markets and Foreign Ownership*, Toronto: Ontario Economic Council
Pavitt, K. (1982) 'R&D, Patenting and Innovative Activities', *Research Policy, 11*, no. 1, 33–51 (February)
Penner, R.G. (1970) 'Policy Reactions and the Benefit of Foreign Investment', *Canadian Journal of Economics*, 3 (May), 213–22
Perry, R. (1982) *The Future of Canada's Auto Industry*, Canadian Institute for Economic Policy, Toronto: James Lorimer & Co.
Pollack, C. and Riedel, J. (1984) *German Firms' Industrial Cooperation With Developing Countries: Status and Prospects*, Paris: OECD
Porter, Michael E. (1980) *Competitive Strategy: Techniques for Analysing Industries and Competitors*, New York: Free Press, Macmillan
Posner, M.V. (1971) 'International Trade and Technical Change', *Oxford Economic Papers* (October)
Prasad, A.J. (1981) 'Licensing as an Alternative to Foreign Investment for Technology Transfer', in R.G. Hawkins and A.J. Prasad (eds.), *Technology Transfer and Economic Development*, Greenwich, Conn.: JAI Press
Richards, Chris F.J. (1985) 'Canadian Direct Investment Position Abroad: Trends and Recent Developments', paper to the Annual Conference of the International Business Council of Canada, mimeo: Statistics Canada:

International Investment Position (April)

Richardson, P.R. (1975) *The Acquisition of New Process Technology by Firms in the Canadian Mineral Industries,* Department of Industry, Trade and Commerce

Robinson, A. (1979) (ed.), *Appropriate Technologies for Third World Development,* London: Macmillan

Robock, S.H. and Simmonds, K. (1983) *International Business and Multinational Enterprises,* 3rd ed., Richard D. Irwin, Homewood

Ronstadt, R. (1977) *Research and Development Abroad by US Multinationals,* New York: Praeger

——— and Kramer, R.J. (1982) 'Getting The Most Out of Innovation Abroad', *Harvard Business Review, 60,* no. 2 (March-April), 94–99

Rugman, Alan M. (1980a) *Multinationals in Canada: Theory, Performance and Economic Impact,* Boston: Martinus Nijhoff

——— (1980b) 'Internalization as a General Theory of Foreign Direct Investment: A Reappraisal of the Literature', *Weltwirtschaftliches Archiv, 116:2,* 365–79

——— (1981) *Inside the Multinationals: The Economics of Internal Markets,* London: Croom Helm, New York: Columbia University Press

——— (1983) 'The Comparative Performance of US and European Multinational Enterprises, 1970–79', *Management International Review, 23:2,* 4–14

——— (1986) *Outward Bound: Canadian Foreign Direct Investment in the United States*, Toronto: C.D. Howe Institute, forthcoming

——— and Eden, L. (1985) (eds.), *Multinationals and Transfer Pricing,* London: Croom Helm

——— and McIlveen, J. (1985) *Megafirms: Strategies for Canada's Multinationals,* Toronto: Methuen

Safarian, A.E. (1966) *Foreign Ownership of Canadian Industry,* Toronto: McGraw Hill

——— (1969) *The Performance of Foreign-Owned Firms in Canada,* Montreal: Canadian American Committee

——— (1983) *Governments and Multinationals: Policies in the Developed Countries,* Toronto: British-North American Committee

Sagafi-Nejad, T., Moxon, R.W. and Perlmutter, H.V. (eds.) (1981) *Controlling International Technology Transfer: Issues, Perspectives, Implications,* New York: Pergamon Press

Saxonhouse, G. (1985) Comments made at the Fourth Annual Workshop on US/Canadian Relations, University of Michigan, Ann Arbor (April)

Scaperlanda, A.E. and Mauer, J.L. (1969) 'The Determinants of US Direct Investment in the EEC', *American Economic Review, 59* (September), 558–68

Science Council of Canada (1979) *Forging the Links: A Technology Policy for Canada,* Ottawa: Supply and Services

Shapiro, C. (1985) 'Patent Licensing and R&D Rivalry', *American Economic Review, 75* (May), 25–30

Shapiro, Daniel M. (1980) *Foreign and Domestic Firms in Canada: A Comparative Study of Financial Structure and Performance,* Toronto: Butterworths

Sharpston, M. (1975) 'International Sub-Contracting', *Oxford Economic*

Papers (March), 94–135
Statistics Canada (1985) *Survey of Intentions on Business Investment Spending*, Ottawa
——— (1985b) *Canadian Imports By Domestic and Foreign Controlled Enterprises*, Ottawa: Minister of Supply and Services
Stobaugh, R. and Telesio, P. (1983) 'Match Manufacturing Policies and Product Strategy', *Harvard Business Review, 61*, no. 2 (March–April), 113–120
Stopford, J.M. and Dunning, J.H. (1983) *Multinationals: Company Performance and Global Trends*, London: Macmillan
Swann, P.L. (1974) 'The International Diffusion of an Innovation', *Journal of Industrial Economics, 22*, 61–9
Teece, D. (1976) *The Multinational Corporation and the Resource Cost of International Transfer*, Cambridge, Mass.: Ballinger Publishers
Telesio, P. (1979) *Technology Licensing and Multinational Enterprises*, New York: Praeger Publishers
Terpstra, V. (1977) 'International Product Policy: The Role of Foreign R&D', *The Columbia Journal of World Business, 12*, no. 4 (Winter), 24–32
——— (1983) *International Marketing*, 3rd ed., Chicago: Dryden Press
Tilton, J. (1971) *International Diffusion of Technology: The Case of Semiconductors*, Washington: Brookings
Toder, E.J., Cordell, N.S. and Burton, E. (1978) *Trade Policy and the US Automobile Industry*, New York, Praeger
Tomlinson, J.W.C. and Thompson, M. (1978) *A Study of Canadian Joint Ventures in Mexico*, Ottawa: Department of Industry, Trade and Commerce
Tsurumi, Y. (1973) 'Japanese Multinational Firms', *Journal of World Trade Law* (February), 74–90
United Nations (1982) *A Survey of the Investment and Licensing Policies of European-based Transnational Bus and Truck Manufacturers in Selected Industrializing Countries*, Joint Unit of the Centre on Transnational Corporations and the Economic Commission for Europe
——— (1983) *Transnational Corporations in the International Auto Industry*, Centre on Transnational Corporations, New York
Vernon, Raymond (1966) 'International Investment and International Trade in the Product Cycle', *Quarterly Journal of Economics, 80* (May), 190–207
——— (1979) 'The Product Cycle Hypothesis in a New International Environment', *Oxford Bulletin of Economics and Statistics* (November), 255–67
——— and Davidson, W.H. (1979) *Foreign Production of Technology-Intensive Products by U.S.-based Multinational Enterprises*, Working Paper, Boston, Graduate School of Business, Harvard University
Wahn Report (1970) *Eleventh Report of the Standing Committee on Defence and External Affairs Respecting Canada-US Relations*, Ottawa: Queen's Printer
Watkins, M. *et al.* (1968) *Foreign Ownership and the Structure of Canadian Industry*, Report of the Task Force on the Structure of Canadian Industry, Ottawa: Queen's Printer

Webley, Simon (1974) *Foreign Direct Investment in the United States: Opportunities and Impediments*, London: British North American Committee

West, E.C. (1971) *Canadian-United States Price and Productivity Differences in Manufacturing Industries, 1963*, Ottawa: Information Canada

Williamson, J. (1983) *The Exchange Rate System*, Institute for International Economics, Policy Analysis No. 5, Cambridge: MIT Press

Williamson, O.E. (1971) 'Vertical Integration of Production: Market Failure Considerations', *American Economic Review, 61*, 2 (May), 112–23

——— (1975) *Markets and Hierarchies: Analysis and Antitrust Implications*, New York: Free Press

——— (1979) 'Transaction-Cost Economics: The Governance of Contractual Relations', *Journal of Law and Economics, 22*, 2 (October), 233–61

Wilson, R.W. (1975) *The Sale of Technology Through Licensing*, New Haven: Yale University

Wonnacott, R. and Wonnacott, P. (1967) *Free Trade Between the United States and Canada: The Potential Economic Effects*, Cambridge, Mass.: Harvard University Press

Wyatt, S. (1984) 'The Role of Patents in Multinational Corporations, Strategies for Growth, Results from Questionnaires', unpublished report prepared for AREPIT, Science Policy Research Unit, University of Sussex (May)

Author Index

Abernathy, W.J. 192–4, 196–7, 199–200
Amsalem, M.A. 84
Anderson, M.L. 194
Aquino, A. 62
AREPIT 100
Armington, P. 32
Arpan, J.S. 122
Astwood, D.M. 57

Bain, J. 193
Baldwin, J.R. 54, 65
Baranson, J. 84
Becker, G. 32
Behrman, J.N. 73, 78, 80, 83–4, 118
Bertin, G.Y. 3
Bhagwati, J. 15, 17
Bishop, P. 149
Blömstrom, M. 84
Bonin, B. 3, 8
Britton, J.M.H. 57, 65
Buckley, P. 58, 84
Business International S.A. 118

Cacnis, D.G. 58
Casson, M. 84
Caves, R.E. 54, 80, 84, 136, 149
Commission of Inquiry (pharmaceuticals) 166
Contractor, F.J. 79–80, 84, 146
Couffin, H. 32
Coughlin, C.C. 118
Cowan, R. 150
Crandall, R.W. 203
Creamer, D. 118
Crookell, H. 79, 84

Daly, D.J. 54–6, 64–5, 67–8, 139, 150
DATAR 173
Davidson, W.H. 75–7, 84, 135–6, 144, 146, 150
Davies, H. 73–4, 78–9, 84
De Bodinat, H. 118
Delapierre, M. 84
DeMelto, D.P. 66, 139, 149
Deslauriers, K. 117
Desranleau, C. 117
Dreze, J. 14
Duerr, M.G. 118
Dunn, R.M. Jr. 121
Dunning, J.H. 8, 11, 58, 84

Eastman, H.C. 5, 31
Economic Council of Canada 139
Erdilek, A. 127
Etemad, H. 4, 117–18
European Management Forum 54

Federal Trade Commission 192, 195–6, 206
Flamm, K. 84
Fortier, Y. 177
Franko, L. 16, 84, 118
Fuss, M. 7, 8, 192, 196–7, 205–6

Globerman, S. 58, 121, 145
Gobez-Ibanez, J.A. 202
Government of Canada 57
Graham, F.D. 13
Gray, H. 122
Grubel, H.B. 32
Guisinger, G. 8

Hakansson, H. 118
Harbour, J.E. 194
Harris, R.G. 55, 68
Helleiner, G.K. 62
Hewitt, G. 118
Hirsch, S. 18
Hufbauer, G. 139

Industry Studies Group 103, 118
INSEE 177
Institut National de la Propriété Industrielle 185–7

Jenkins, G.P. 69
Jones, R.W. 178

Katrak, H. 32
Killing, J.P. 81–2, 149
Kojima, K. 11
Kopits, G.F. 80
Krueger, A.O. 15
Kwoka, J. 206

Lafleur, B. 57
Lake, A.W. 118
Lancaster, K. 14
Lassudrie-Duchene, B. 13
Linder, S.B. 15

MacCharles, D.C. 2, 6, 58–9, 61–5
Machlup, F. 58–9
McFetridge, D.G. 5, 8, 139–40, 148, 150
McIlveen, J. 134
McMullen, K. 147
Maddala, G.S. 150
Madeuf, B. 2, 36
Magee, S.P. 32
Mansfield, E. 84, 111, 118, 136, 138, 140–1, 145
Mathewson, G.F. 80
Messerlin, P.A. 6, 175
Muchielli, J-L. 1, 8, 31–2
Mundell, R.A. 178

Nalin, C. 26
Nasbeth, L. 139, 141, 150
Neary, J.P. 178

OECD 44, 46, 52, 77
Ohlin, B. 32
Oman, C. 8, 84
Onida, F. 84
Ozawa, T. 84

Palda, K.S. 54, 58, 66
Parry, T.G. 84
Pattison, J.C. 121
Pavitt, K. 103
Perry, R. 192, 195
Pollack, C. 84
Porter, M.E. 8, 129
Posner, M.V. 13

Robinson, A. 84
Robock, S.H. 84, 118
Ronstadt, R. 118
Rugman, A.M. 4, 33, 57–8, 121, 124, 128, 130, 134

Safarian, A.E. 8, 57–8, 63, 84, 121
Sagafi-Nejad, T. 84
Saxonhouse, G. 202
Scaperlanda, A.E. 18
Shapiro, C. 136
Shapiro, M. 121
Sharpston, M. 32
Soulié, D. 6
Statistics Canada 57, 60, 123, 125
Stobaugh, R. 118
Stopford, J.M. 102–3, 118
Swann, P.L. 141, 150
Swedenborg, B. 118

Teece, D. 80–1, 83
Telesio, P. 73–4, 76–8, 84
Terpstra, V. 118
Tilton, J. 139
Toder, E.J. 193, 206
Tsurumi, Y. 26

US Department of Commerce 126

Vernon, R. 13, 26, 136–7
Verreault, R. 84

Wahn Report 122
Watkins, M. 122
Webley, S. 122
West, E.C. 55
Williamson, O.E. 8, 146, 149
Wilson, R.W. 73, 84
Wonnacott, R. 55
Wyatt, S. 118

Zayat, H. 117

Subject Index

automobile industry 14, 180–90 *passim*, 191–205 *passim*

barriers to trade
- and corporate decision-making 53–4
- and FDI 57, 124, 128, 167–8, 174
- and licensing 74
- and productivity 55
- and technological transfer order 143
- costs of 178–9
- in France 167–78 *passim*
- non-tariff barriers 168–71
- voluntary export restrictions 169, 176, 206n8

comparative advantage
- and FDI 11–33 *passim*
- Canada 55, 64–5
- Canadian MNEs 131–3
- engineering of 137–9
- France 175, 188
- Japan 168, 192
- R&D intensity 50–1
- sectoral stratification 21–2
- *see also* productivity

competitiveness
- and knowledge transfer 53
- technological 40
- *see also* comparative advantage productivity

competitive strategy
- automotive equipment producers 188–9
- centralisation of R&D 116–17
- MNE 19–21, 90–9 *passim*, 127–9, 131–3
- pharmaceutical firms 163–4
- restrictive clauses 79, 82–3, 149
- technological competitiveness 51, 86–9, 98–9, 145–7
- *see also* internalisation

core technology 44, 76–7, 98–9

demand
- and R&D profits 155
- and transfer order 143–4
- FDI 17–18, 22, 124
- foreign trade 13
- US automobile industry 203

Directory of multinational enterprises 102–3

diversification
- and FDI 127
- and licensing 78
- horizontal 54
- vertical 54
- *see also* core technology

economies of scale
- and FDI 18, 127
- and international trade 13–14
- and joint ventures 191
- Canada 54–5
- centralisation of R&D 111
- Japan 193
- pharmaceutical industry 153, 157–8

electronics industry 167–79 *passim*

factor endowments
- and factor imports 178
- and FDI 17, 21, 27–31
- and international trade 13
- Japan 202–4
- technology transfer 141

factor market distortions
- and FDI 19
- and international trade 15
- technical change 68–9

Foreign Investment Review Agency 127, 148

Heckscher-Ohlin model 13–14

internalisation 73–81 *passim*, 86–9, 96–9, 128–9, 145–8
international trade
  Canada 53–69 *passim*
  determinants 11–16
  France 167–79 *passim*
  MNEs 60–2
  technological performance 181–2

joint ventures 78–82, 99, 198–205

Kanban system 202–4

licensing
  advantages of 76
  and barriers to entry 74
  and competition 146
  compulsory 161–6 *passim*
  cross-licensing 78, 89, 94–5
  diversification 78
  market structure 73–4, 75n4
  oligopoly 73–4
  pricing 79–80
  R&D 74–6, 78, 81–2
  see also barriers to trade, patenting, product market distortions, R&D, technology transfer

Multinational enterprise database 136–42, 146–7

organisational structure
  and knowledge transfer 67–9
  centralisation of R&D 111
  Japanese automobile industry 201–3
patenting
  activity 90–1, 102, 107–17 *passim*
  and licensing 89–94 *passim*
  British Patent Act 161
  Canadian Patent Act 160–2
  competitive strategy 163–4
  compulsory licensing 161–6 *passim*
  French automobile industry 184–7
  product differentiation 156–9
  protection 153–62 *passim*
  see also licensing
pharmaceutical industry 153–66 *passim*
political risk and FDI 127–8, 133
product cycle 24–6, 136–9
product market distortions
  and FDI 18–19
  and international trade 14–15
  licensing 75, 79, 108–9
  R&D 155
productivity
  and free trade 54–5
  capacity utilisation 195–6
  Japanese automobile industry 191–206 *passim*
  MNEs 63–5
  specialisation in Canada 68–9

R&D (research and development)
  and FDI 57–9, 85–7
  and licensing 74–82 *passim*
  by subsidiaries in Canada 65–6
  centralisation of 102–17 *passim*
  compulsory licensing 165
  duplication of 154–6
  French industry 174, 181
  high-intensity industries 44–7
  optimal 155
  patent protection 89–94 *passim*
  pharmaceutical industry 156, 158–9
  propensity to export 47–9
  transfer lags 141–2
  see also demand, economies of scale
Ricardian model 13

specialisation
  and R&D intensity 48–5
  by subsidiaries 59, 63–4

inappropriate 178
internalisation 86–7, 96
intra-industry trade 69
productivity gains 68

technology
appropriate 26–7, 189, 192, 203
balance of payments 39–52 *passim*, 138–9
dependence 39–44
differences; and international trade 12–13; and FDI 16–17, 21–7 *passim*; Japan and US automobile industry 198, 201–3; production 34–5, 101–5
technology transfer
age of technology 136–8
and barriers to trade 193–5
Canada 60, 65–6
costs incurred 80–1, 143
country positions 39, 50–1, 140–2
environmental conditions 88–9
indicators of 36–9, 52
Japan to France and US 176–8, 183–9, 191–2
lags 136–47 *passim*
partners 51, 86, 90–100 *passim*, 145–7
USA to France and Canada 140–2
see also internalisation, joint ventures, licensing